CITES Aloe and Pachypodium Checklist

Edited by: Urs Eggli

Authors: Leonard E Newton (Aloe)
             Gordon D Rowley (Pachypodium)

First published in 2001

General editor of series: Jacqueline A Roberts

ISBN 1 84246 034 X

Produced with the financial assistance of:

The CITES Nomenclature Committee
Sukkulenten-Sammlung Zürich
The Royal Botanic Gardens, Kew

Cover design by Media Resources RBG Kew
Printed in Great Britain by The Cromwell Press

CONTENTS

Preamble

TABLE DES MATIERES

Préambule

ÍNDICE

Preámbulo

CITES CHECKLIST - ALOE AND PACHYPODIUM

PREAMBLE

1. Background

The 1992 Conference of the Parties to the Convention on International Trade in Endangered Species of Wild Fauna and Flora (CITES) adopted the *CITES Cactaceae Checklist* as the guideline when making reference to the species of the genera concerned. This was the first CITES plant checklist, which was followed by the publication of the *CITES Orchid Checklist Volume 1*, in 1995.

These references have proved to be an important tool in the day to day implementation of CITES for plant species. The combination of support from the CITES Conferences of the Parties, individual party States, scientific institutions and organisations has facilitated the preparation and publication of the *CITES Cactaceae Checklist (second edition)*, the *CITES Orchid Checklist Volume 2*, the *CITES Checklist of succulent Euphorbia taxa*, and the *CITES Bulb Checklist*. All of these works have been adopted by the Conference of the Parties, as the guidelines when making reference to the accepted names of the genera concerned.

This checklist is the result of co-operation between the Städtische Sukkulenten-Sammlung (Switzerland) and the Royal Botanic Gardens, Kew (UK).

This taxonomic list is an extract from the database of *Repertorium Plantarum Succulentarum* (RPS) data maintained at the Municipal Succulent Collection Zürich. The database contains nomenclatural and taxonomic information published in serialized form in the yearly issues of *Repertorium Plantarum Succulentarum*, initiated and currently published by the *International Organization for Succulent Plants Study* (IOS). Where possible, additional taxa have been included (see IOS Bull. 5(4): 172-173, 1992). (Source: ZSS & IOS 2001. *Notes on this Extract from the Repertorium Plantarum Succulentarum Database*).

This checklist incorporates the largest collection of published synonyms for these genera. Further data, revisions, edits, and updates have been added during the compilation of this list. This checklist is therefore up-to-date and should be useful for a number of years.

2. How to use the Checklist

It is intended that this Checklist be used as a quick reference for checking accepted names, synonymy and distribution. The reference is therefore divided into three main parts for ease of use by non-botanists/taxonomists. Part IV has been included to provide more information for users with a greater understanding of plant nomenclature:

Part I: All Names in Current Use

An alphabetical list of all accepted names and synonyms included in this checklist – a total of 1026 names (578 accepted and 448 synonyms)

Part II: Accepted Names in Current Use

Separate lists for each genus. Each list is ordered alphabetically by the accepted name and details are given on current synonyms and distribution.

Preamble

Part III: Country Checklist
Accepted names from all genera included in this checklist are ordered alphabetically under country of distribution.

Part IV: Accepted Taxa
Separate lists for each genus including further details on homotypic names, basionyms, heterotypic synonyms etc.

3. **Conventions employed in Parts I, II and III**
   a) Accepted names are presented in **bold roman** type.
      Synonyms are presented in *italic* type.

   b) Duplicate names:

      In Part I, the author's name appears after each taxon where the taxon name appears twice or more (unless the author's name is the same) e.g., *Aloe abyssinica* Hook.f., *Aloe abyssinica* Salm-Dyck and *Aloe abyssinica* A.Berger

      It is also necessary to double-check by reference to the distribution as detailed in Part II. For example, if the name given was 'Aloe angustifolia' and it was known that the plant in question came from Angola, this would indicate that the species was **Aloe zebrina**, being traded under the synonym *Aloe angustifolia* - **Aloe angustifolia** is only found in South Africa.

   c) Natural hybrids have been included in the checklist and are indicated by the multiplication sign ×. They are arranged alphabetically within the lists.

4. **Number of names included for each genus:**
Aloe (Accepted: 548, Synonyms: 428), Pachypodium (Accepted: 30, Synonyms: 20)

5. **Geographical areas**
Country names follow the United Nations standard as laid down in *Country Names. Terminology Bulletin* No. 347/Rev. 1. 1997, United Nations.

6. **Aloes and Pachypodiums controlled by CITES**

Appendix II:

Aloe spp. *-116 #1
Pachypodium spp. *#1

Appendix I:

Aloe albida
Aloe albiflora
Aloe alfredii
Aloe bakeri
Aloe bellatula
Aloe calcairophila
Aloe compressa = 460
Aloe delphinensis
Aloe descoingsii
Aloe fragilis

Aloe haworthioides = 461
Aloe helenae
Aloe laeta = 462
Aloe parallelifolia
Aloe parvula
Aloe pillansii
Aloe polyphylla
Aloe rauhii
Aloe suzannae
Aloe thorncroftii
Aloe versicolor
Aloe vossii

Pachypodium ambongense
Pachypodium baronii
Pachypodium decaryi

Interpretation:

The symbol (#) followed by a number placed against the name of a species or higher taxon included in Appendix II designates parts or derivatives that are specified in relation thereto for the purposes of the Convention as follows:

#1   Designates all parts and derivatives, except:

a) seeds, spores and pollen (including pollinia);

b) seedling or tissue cultures obtained *in vitro*, in solid or liquid media, transported in sterile containers; and

c) cut flowers of artificially propagated plants.

The symbol (-) followed by a number placed against the name of a species or higher taxon denotes that designated geographically separate populations, species, groups of species or families of that species or taxon are excluded from the appendix concerned, as follows:

- 116   *Aloe vera*; also referenced as *Aloe barbadensis*.

The symbol (=) followed by a number placed against the name of a species, subspecies or higher taxon denotes that the name of that species, subspecies or taxon shall be interpreted as follows:

= 460   Includes *Aloe compressa* var. *rugosquamosa* and *Aloe compressa* var. *schistophila*

= 461   Includes *Aloe haworthioides* var. *aurantiaca*

= 462   Includes *Aloe laeta* var. *maniaensis*

7. Abbreviations, botanical terms, and Latin

Not all these abbreviations, botanical terms and Latin will appear in this Checklist; however, they have been included as a useful reference.

Note: words in *italics* are Latin

ambiguous name a name which has been applied to different taxa by different authors, so that it has become a source of ambiguity

anon. anonymous; without author or author unknown

auct. *auctorum*: of authors

CITES Convention on International Trade in Endangered Species of Wild Fauna and Flora

cultivar an individual, or assemblage of plants maintaining the same distinguishing features, which has been produced or is maintained (propagated) in cultivation

cultivation the raising of plants by horticulture or gardening; not immediately taken from the wild

descr. *descriptio*: the description of a species or other taxonomic unit

distribution where plants are found (geographical)

ed. editor

edn. edition (book or journal)

eds. editors

epithet the last word of a species, subspecies, or variety (etc.), for example: *volkensii* is the species epithet for the species *Aloe volkensii* and *multicaulis* is the subspecific epithet for *Aloe volkensii* ssp. *multicaulis*

escape a plant that has left the boundaries of cultivation (e.g. a garden) and is found occurring in natural vegetation

ex *ex*: after; may be used between the name of two authors, the second of whom validly published the name indicated or suggested by the first

excl. *exclusus*: excluded

forma *forma*: a taxonomic unit inferior to variety

hort. *hortorum*: of gardens (horticulture); raised or found in gardens; not a plant of the wild

ICBN International Code for Botanical Nomenclature

in prep. in preparation

in sched. *in scheda*: on a herbarium specimen or label

in syn. *in synonymia*: in synonymy

incl. including

ined. *ineditus*: unpublished

introduction a plant which occurs in a country, or any other locality, due to human influence (by purpose or chance); any plant which is not native

key a written system used for the identification of organisms (e.g. plants)

leg. *legit*: he gathered; the collector

misspelling a name that has been incorrectly spelt; not a new or different name

morphology the form and structure of an organism (e.g. a plant)

name causing confusion a name that is not used because it cannot be assigned unambiguously to a particular taxon (e.g. a species of plant)

native an organism (e.g. a plant) that occurs naturally in a country, or region, etc.

naturalized a plant which has either been introduced (see introduction) or has escaped (see escape) but which looks like a wild plant and is capable of reproduction in its new environment

nom. ambig. *nomen ambiguum*: ambiguous name

nom. cons. prop. *nomen conservandum propositum*: name proposed for conservation under the rules of the International Code for Botanical Nomenclature (ICBN)

nom. illeg *nomen illegitimum*: illegitimate name
nom. *nomen*: name
nom. nud. *nomen nudum*: name published without description
nomenclature branch of science concerned with the naming of organisms (e.g. plants)
non *non*: not
only known from cultivation a plant which does not occur in the wild, only in cultivation
orthographic variant an alternative spelling for the same name
p.p. *pro parte*: partly, in part
pro parte *pro parte*: partly, in part
provisional name name given in anticipation of a valid description
sens. lat. *sensu lato*: in the broad sense; a taxon (usually a species) and all its subordinate taxa (e.g. subspecies) and/or other taxa sometimes considered as distinct
sens. *sensu*: in the sense of; the manner in which an author interpreted or used a name
sensu *sensu*: in the sense of; the manner in which an author interpreted or used a name
sic *sic*, used after a word that looks wrong or absurd, to show that it has been quoted correctly
spp. species
ssp. subspecies
synonym a name that is applied to a taxon but which cannot be used because it is not the accepted name – the synonym or synonyms form the synonymy
taxa plural of taxon
taxon a named unit of classification, e.g. genus, species, subspecies
var. variety

*thanks to Dr Aaron Davis, RBG Kew, for the provision of this guide

8. Bibliography

Brummitt, R.K. and Powell, C.E. (Eds.) (1992). *Authors of Plant Names.* Royal Botanic Gardens, Kew (UK).

Newton, L.E. (2001). Sansevieria. In: U.Eggli (ed.), *Illustrated Handbook of Succulent Plants: Monocotyledons:* 261-272. Springer, Berlin.

Rapanarivo, S.H.J.V, Lavranos, J.J., Leeuwenberg, A.J.M. & Röösli, W. (1999). *Pachypodium (Apocynaceae), Taxonomy, habitats & cultivation.* A.A. Balkema, Rotterdam / Brookfield.

Rowley, G.D. (2002). Pachypodium (Apocynaceae). In U. Eggli (ed.), *Illustrated Handbook of Succulent Plants: Dicotyledons* (in press). Springer, Heidelberg.

Terminology Bulletin No. 347/Rev. 1. 1997, United Nations.

LISTE DES *ALOE* ET *PACHYPODIUM* CITES

PREAMBULE

1. Historique

En 1992, la Conférence des Parties à la Convention sur le commerce international des espèces de faune et de flore sauvages menacées d'extinction (CITES) a adopté la *CITES Cactaceae Checklist* (liste des Cactacées CITES) comme référence aux espèces de cette famille. C'était la première liste de plantes couvertes par la CITES; elle a été suivie, en 1995, de la *CITES Orchid Checklist* (liste des Orchidées CITES), *Volume 1*.

Ces références se sont révélées très utiles pour la mise en œuvre pratique de la Convention. L'appui combiné de la Conférence des Parties à la CITES, de Parties individuelles, de certaines institutions et d'organisations scientifiques (IOS, par exemple) a facilité la préparation et la publication de la *CITES Cactaceae Checklist (deuxième édition)*, de la *CITES Orchid Checklist Volume 2*, de la *CITES Checklist of Succulent Euphorbia Taxa* et de la *CITES Bulb Checklist*. Toutes ont été adoptées par la 10ᵉ session de la Conférence des Parties comme listes normalisées de référence pour les noms acceptés des taxons concernés.

La présente liste taxonomique résulte de la coopération des *Städtische Sukkulenten-Sammlung* (Suisse) et du *Royal Botanic Gardens, Kew*, Royaume-Uni (Jardins botaniques royaux, JBR).

Cette liste est tirée de la base de données du *Repertorium Plantarum Succulentarum* (RPS) de la Collection municipale de plantes grasses de Zürich. Cette base de données contient des informations sur la nomenclature et la taxonomie publiées par séries dans les numéros annuels du *Repertorium Plantarum Succulentaraum*, lancé et actuellement publié par l'*International Organization for Succulent Plants Study* (IOS). Des taxons supplémentaires ont été inclus quand c'était possible (voir IOS Bull. 5(4): 172-173, 1992. (Source: ZSS & IOS 201. *Notes on this Extract from the Repertorium Plantarum Succulentarum Database*).

La Liste inclut le plus grand nombre de synonymes publiés pour ces trois genres, ainsi que d'autres données, révisions et corrections; elle est donc à jour et devrait être utile pendant plusieurs années.

2. Comment utiliser la Liste?

Cette liste peut être utilisée comme référence pour vérifier rapidement les noms acceptés, les synonymes et la répartition géographique. La référence est donc divisée en trois grandes parties pour en faciliter l'utilisation par les non-botanistes/taxonomistes. La quatrième partie a été incluse pour que l'utilisateur ait davantage d'informations et une meilleure compréhension de la nomenclature des plantes:

Première partie: Tous les noms d'usage courant
Liste alphabétique de tous les noms reconnus et des synonymes inclus dans la Liste - au total 1026 noms (578 noms reconnus et 448 synonymes).

Deuxième partie: Noms acceptés d'usage courant
Listes séparées pour chaque genre. Chaque liste est donnée dans l'ordre alphabétique des noms reconnus et comporte des indications sur les synonymes et la répartition géographique actuels.

Preambule

Troisième partie: Liste des pays
Les noms reconnus de tous les genres inclus dans cette liste sont donnés par ordre alphabétique sous chaque pays de l'aire de répartition.

Quatrième partie: Taxons acceptés
Autre indications de noms homotypiques, de basionymes et de synonymes hétérotypiques.

3. **Conventions employées dans les première, deuxième et troisième parties**
   a) Les noms acceptés sont imprimés **en gras**
      les synonymes *en italique.*

   b) Noms identiques pour des taxons différents:

      Dans la première partie, le nom de l'auteur est indiqué après chaque taxon lorsque le nom du taxon apparaît deux fois ou plus (à moins que le nom de l'auteur soit le même). Exemple: *Aloe abyssinica* Hook.f., *Aloe abyssinica* Salm-Dyck et *aloe abyssinica* A.Berger

      Il convient aussi de faire une double vérification en se référant à la répartition géographique détaillée dans la deuxième partie. Exemple: si le nom donné est "Aloe angustifolia" et si l'on sait que la plante en question vient de l'Angola, cela signifie que l'espèce est **Aloe zebrina**, commercialisée sous le synonyme *Aloe angustifolia* - **Aloe angustifolia** ne se trouve qu'en Afrique du Sud.

   c) Une sélection d'hybrides a été incluse dans la liste. Ils sont signalés par le signe de multiplication ×. Ils sont placés par ordre alphabétique dans les première, deuxième et troisième parties.

4. **Décompte des noms retenus pour chaque genre**
Aloe (548 acceptés, 428 synonymes); *Pachypodium* (30 acceptés, 20 synonymes).

5. **Régions**
Les noms des pays sont ceux figurant dans le *Country Names. Terminology Bulletin* No. 347/Rev. 1. 1997, United Nations.

6. **Aloès et pachypodiums soumis aux contrôles CITES**

Annexe II:

Aloe spp. *-116 #1
Pachypodium spp. *#1

Annexe I:

Aloe albida
Aloe albiflora
Aloe alfredii
Aloe bakeri
Aloe bellatula
Aloe calcairophila
Aloe compressa = 460
Aloe delphinensis

Aloe descoingsii
Aloe fragilis
Aloe haworthioides = 461
Aloe helenae
Aloe laeta = 462
Aloe parallelifolia
Aloe parvula
Aloe pillansii
Aloe polyphylla
Aloe rauhii
Aloe suzannae
Aloe thorncroftii
Aloe versicolor
Aloe vossii

Pachypodium ambongense
Pachypodium baronii
Pachypodium decaryi

Interprétation:

Le signe (#) suivi d'un nombre placé après le nom d'une espèce ou d'un taxon supérieur inscrit à l'Annexe II sert à désigner des parties ou produits obtenus à partir de ladite espèce ou dudit taxon et qui sont mentionnés comme suit aux fins de la Convention:

#1   Sert à désigner toutes les parties et tous les produits, sauf:

   a) les graines, les spores et le pollen (y compris les pollinies);

   b) les cultures de plantules ou de tissus obtenues *in vitro*, en milieu solide ou liquide, et transportées en conteneurs stériles; et

   c) les fleurs coupées des plantes reproduites artificiellement.

Le signe (-) suivi d'un nombre placé après le nom d'une espèce ou d'un taxon supérieur signifie que des populations géographiquement isolées, espèces, groupes d'espèces ou familles, de ladite espèce ou dudit taxon, sont exclus de l'annexe en question, comme suit:

- 116   *Aloe vera*; aussi appelé *Aloe barbadensis*.

Le signe (=) suivi d'un nombre placé après le nom d'une espèce, d'une sous-espèce ou d'un taxon supérieur signifie que la dénomination de ladite espèce ou sous-espèce ou dudit taxon doit être interprétée comme suit:

= 460   Comprend *Aloe compressa* var. *rugosquamosa* et *Aloe compressa* var. *schistophila*

= 461   Comprend *Aloe haworthioides* var. *aurantiaca*

= 462   Comprend *Aloe laeta* var. *maniaensis*

7. Abréviations, termes botaniques, et mots latins

Ces termes de botanique, noms latins et abréviations ne sont pas tous utilisés dans la Liste. Ils sont inclus pour référence.

Note: les mots *en italique* sont d'origine latine

ambiguous name (nom ambigu) nom donné à différents taxons par différents auteurs, ce qui crée une ambiguïté
anon. anonyme; sans auteur
auct. *auctorum*: d'auteurs
CITES Convention sur le commerce international des espèces de faune et de flore sauvages menacées d'extinction
cultivar spécimen ou groupe de plantes conservant les mêmes caractéristiques distinctives, produites ou conservées (propagées) en culture
cultivation (culture) obtention de plantes par horticulture ou jardinage, par opposition au prélèvement dans la nature
descr. *descriptio* description d'une espèce ou d'une autre entité taxonomique
distribution (aire de répartition géographique) région(s) où se trouve les plantes
ed. éditeur
edn. édition (d'un livre ou d'un périodique)
eds. éditeurs
epithet (épithète) dernier mot d'une espèce, d'une sous-espèce ou d'une variété (etc.). Exemple: *volkensii* est l'épithète de l'espèce *Aloe volkensii* et *multicaulis* l'épithète infraspécifique de *Aloe volkensii* ssp. *multicaulis*
escape (échappée) qualifie une plante qui a quitté l'enceinte de culture (jardin, par exemple) et qu'on retrouve dans la végétation naturelle
ex *ex* d'après; peut être utilisé entre deux noms d'auteurs, dont le second a validement publié le nom d'après les indications ou suggestions du premier
excl. *exclusus* exclu
hort. *hortorum* de jardins (horticole); plante cultivée ou se trouvant dans des jardins horticoles, par opposition à une plante d'origine sauvage
ICBN (CINB) Code international de la nomenclature botanique
in prep. en préparation
in sched. *in scheda* sur un spécimen d'herbier ou une étiquette
in syn. *in synonymia* en synonymie
incl. incluant
ined. *ineditus* non publié
introduction résultat d'une activité humaine (volontaire ou non) aboutissant à ce qu'une plante non indigène se retrouve dans un pays ou une région
key (clé) système écrit utilisé pour la détermination d'organismes (plantes, par exemple)
leg. *legit* il ramassa; le collecteur
misspelling (faute d'orthographe) nom mal orthographié, par opposition à un nom nouveau ou différent
morphology (morphologie) forme et structure d'un organisme (d'une plante, par exemple)
name causing confusion (nom causant une confusion) nom qui n'est pas utilisé parce qu'il ne peut être assigné sans ambiguïté à un taxon particulier (à une espèce de plante, par exemple)
native (indigène) qualifie un organisme (une plante, par exemple) prospérant naturellement dans un pays ou une région etc.
naturalized (naturalisée) qualifie une plante introduite (voir introduction) ou échappée (voir échappée) qui ressemble à une plante sauvage et qui se propage dans son nouvel environnement

nom. *nomen* nom
nom. ambig. *nomen ambiguum* nom ambigu
nom. cons. prop. *nomen conservandum propositum* nom dont le maintien a été proposé d'après les règles du *International Code of Botanical Nomenclature* (Code international de la nomenclature botanique)
nomenclature branche de la science qui nomme les organismes (les plantes, par exemple)
non *non* pas
only known from cultivation (connue seulement en culture) qualifie une plante qu'on ne trouve pas à l'état sauvage
orthographic variant (variante orthographique) même nom orthographié différemment
pro parte *pro parte* partiellement, en partie
provisional name (nom provisoire) nom donné par anticipation d'une description
sens. *sensu* au sens de; manière dont un auteur interprète ou utilise un nom
sens. lat. *sensu lato* au sens large; un taxon (habituellement une espèce) et tous ses taxons inférieurs (sous-espèce, etc.) et/ou d'autres taxons parfois considérés comme distincts
sic *sic*, utilisé après un mot qui semble faux ou absurde; indique que ce mot est cité textuellement
synonym (synonyme) nom donné à un taxon mais qui ne peut être utilisé parce que ce n'est pas le nom accepté; le ou les synonymes forment la synonymie
taxa pluriel de taxon
taxon unité taxonomique à laquelle on a attribué un nom - genre, espèce, sous-espèce, etc.
var. variété

* Nous remercions M. Aaron Davis, des JBR de Kew, d'avoir fourni ce guide

8. Bibliographie

Brummitt, R.K. and Powell, C.E. (Eds.) (1992). *Authors of Plant Names*. Royal Botanic Gardens, Kew (UK).

Newton, L.E. (2001). Sansevieria. In: U.Eggli (ed.), *Illustrated Handbook of Succulent Plants: Monocotyledons:* 261-272. Springer, Berlin.

Rapanarivo, S.H.J.V, Lavranos, J.J., Leeuwenberg, A.J.M. & Röösli, W. (1999). *Pachypodium (Apocynaceae), Taxonomy, habitats & cultivation*. A.A. Balkema, Rotterdam / Brookfield.

Rowley, G.D. (2002). Pachypodium (Apocynaceae). In U. Eggli (ed.), *Illustrated Handbook of Succulent Plants: Dicotyledons* (in press). Springer, Heidelberg.

Terminology Bulletin No. 347/Rev. 1. 1997, United Nations.

LISTA CITES - ALOE Y PACHYPODIUM

PREÁMBULO

1. Antecedentes
En 1992, la Conferencia de las Partes en la Convención sobre el Comercio Internacional de Especies Amenazadas de Fauna y Flora Silvestres (CITES) adoptó la *CITES Cactaceae Checklist* como obra de referencia al hacer alusión a las especies de la familia en cuestión. Se trataba de la primera lista de plantas CITES, que fue seguida por la *CITES Orchid Checklist Volume 1*, en 1995.

Se ha puesto de manifiesto que estas referencias son un valioso instrumento en las tareas diarias de aplicación de la CITES para las especies de plantas. El apoyo de la Conferencia de las Partes sumado al de los Estados Partes, las instituciones científicas y las organizaciones han hecho posible la preparación y publicación de la *CITES Cactaceae Checklist (second edition)*, la *CITES Orchid Checklist Volume 2*, la *CITES Checklist of succulent Euphorbia taxa* y la *CITES Bulb Checklist*. Todas estas listas han sido adoptadas por la décima reunión de la Conferencia de las Partes como obras de referencia al hacer alusión a los nombres aceptados de los taxa de que trata.

La presente lista es el resultado de la cooperación entre el Städtische Sukkulenten-Sammlung (Suiza) y el Real Jardín Botánico de Kew (Reino Unido).

Esta lista taxonómica es un extracto de la base de datos *Repertorium Plantarum Succulentarum* (RPS), mantenida por la Colección Municipal de Suculentas de Zürich. En la base de datos figura información sobre nomenclatura y taxonomía publicada en forma de entregas en las publicaciones anuales de *Repertorium Plantarum Succulentarum*, iniciada y actualmente publicada por la Organización Internacional para el Estudio de Plantas Suculentas (IOS). En la medida de lo posible, se han incluido taxa adicionales (véase IOS Bull. 5(4): 172-173, 1992). (Fuente: ZSS & IOS 2001. *Notes on this Extract from the Repertorium Plantarum Succulentarum Database*).

Es esta lista se incorpora el mayor número de sinónimos publicados para los tres géneros. Durante la compilación de la misma se añadieron nuevos datos, revisiones, ediciones y actualizaciones. Por ende, esta lista está actualizada y debería ser útil durante varios años.

2. ¿Cómo utilizar esta lista?

La finalidad consiste en que esta lista se utilice como referencia rápida para verificar los nombres aceptados, los sinónimos y la distribución. En consecuencia, se ha dividido en tres partes para facilitar su utilización por aquellas personas que no sean botánicos/taxonomistas. Se ha incluido la Parte IV para ofrecer mayor información a los usuarios con una mayor comprensión de la nomenclatura de las plantas:

Parte I: Todos los nombres utilizados normalmente
Una lista por orden alfabético de todos los nombres y sinónimos aceptados - – un total de 1026 nombres (578 aceptados y 448 sinónimos)

Parte II: Nombres aceptados utilizados normalmente
Listas separadas para cada género. En cada lista se presentan por orden alfabético los nombres aceptados, con información sobre los sinónimos y la distribución.

Parte III: Lista por países
Los nombres aceptados para todos los géneros incluidos en esta lista se presentan por orden alfabético según el país de distribución.

Parte IV: Taxa aceptados
Mayor información sobre nombres homotípicos, basionímicos y sinónimos heterotípicos.

3. **Sistema de presentación utilizado en las Partes I, II y III**
   a) Los nombres aceptados se presentan en **negrita**.
      Los sinónimos se presentan en letra *cursiva*.

   b) Nombres duplicados:

      En la Parte I, el nombre del autor aparece después de cada taxón, cuando el nombre de dicho taxón aparece más de una vez (al menos que el nombre del autor sea el mismo) p.e., *Aloe abyssinica* Hook.f., *Aloe abyssinica* Salm-Dyck y *Aloe abyssinica* A.Berger

      Es preciso también efectuar doble verificación en relación con la distribución, tal como se detalla en la Parte II. Por ejemplo, si el nombre dado es 'Aloe angustifolia' y se sabe que la planta en cuestión procede de Angola, esto significa que se trata de la especie **Aloe zebrina**, comercializada bajo el sinónimo de *Aloe angustifolia* - **Aloe angustifolia** sólo se encuentra en Sudáfrica.

   c) En la lista se han incluido los híbridos naturales y se indican con el signo de multiplicar "×". Se presentan por orden alfabético.

4. **Número de nombres incluidos para cada género:**
   Aloe (Aceptados: 548, Sinónimos: 428), Pachypodium (Aceptados: 30, Sinónimos: 20)

5. **Áreas geográficas**
   Para los nombres de los países se ha seguido la referencia oficial de las Naciones Unidas. *Terminology Bulletin* No. 347/Rev. 1, 1997, United Nations.

6. **Aloes y Pachypodiums amparadas por la CITES**

Apéndice II:

Aloe spp. *-116 #1
Pachypodium spp. *#1

Apéndice I:

Aloe albida
Aloe albiflora
Aloe alfredii
Aloe bakeri
Aloe bellatula
Aloe calcairophila
Aloe compressa = 460
Aloe delphinensis
Aloe descoingsii

Aloe fragilis
Aloe haworthioides = 461
Aloe helenae
Aloe laeta = 462
Aloe parallelifolia
Aloe parvula
Aloe pillansii
Aloe polyphylla
Aloe rauhii
Aloe suzannae
Aloe thorncroftii
Aloe versicolor
Aloe vossii

Pachypodium ambongense
Pachypodium baronii
Pachypodium decaryi

Interpretación:

El signo (#) seguido de un número colocado junto al nombre de una especie o de un taxón superior incluido en el Apéndice II designa las partes o derivados que están en estrecha relación con la misma a los efectos de la Convención, a saber:

#1   Designa todas las partes y derivados, excepto:

a) las semillas, las esporas y el polen (inclusive las polinias);

b) los cultivos de plántulas o de tejidos obtenidos *in vitro*, en medios sólidos o líquidos, que se transportan en envases estériles; y

c) las flores cortadas de plantas reproducidas artificialmente.

El signo (-) seguido de un número colocado junto al nombre de una especie o de un taxón superior significa que algunas poblaciones geográficamente aisladas, especies, grupos de especies o familias de dicha especie o de dicho taxón están excluidas del Apéndice concernido, como sigue:

- 116   *Aloe vera*; también citada como *Aloe barbadensis*.

El signo (=) seguido de un número colocado junto al nombre de una especie, una subespecie o un taxón superior significa que la denominación de la especie, subespecie o taxón superior deber ser interpretada como sigue:

= 460   Incluye *Aloe compressa* var. *rugosquamosa* y *Aloe compressa* var. *schistophila*

= 461   Incluye *Aloe haworthioides* var. *aurantiaca*

= 462   Incluye *Aloe laeta* var. *maniaensis*

7. Abreviaciones, términos botánicos y expresiones latinas*

En esta Lista no aparecen todas las abreviaturas, términos botánicos y en latín, pese a que se han incluido como referencia útil.

Nota: las expresiones latinas aparecen en *cursiva*

ambiguous name (nombre ambiguo) un nombre utilizado por distintos autores para diferentes taxa, de manera que da motivo a confusión

anon. Anonymous; autor desconocido

auct. *auctorum* de autores

CITES Convención sobre el Comercio Internacionale de Especies Amenazadas de Fauna y Flora Silvestres

cultivation (cultivo) el cultivo de plantas mediante horticultura o jardinería; no se ha recolectado inmediatamente del medio silvestre

cultivar un ejemplar, o una agrupación de plantas, que tiene los mismos rasgos característicos, que ha sido producido o se mantiene (reproduce) en cultivo

descr. *descriptio* la descripción de una especie o de otra unidad taxonómica

distribution (distribución) donde se encuentran las plantas (geográfica)

ed. editor

edn. edición (libro o revista)

eds. editores

eptithet (epíteto) la última palabra de una especie, subespecie o variedad (etc.), por ejemplo: *lutea* es el epíteto de la especie *Sternbergia lutea* y *byzantinus* es el epíteto subespecífico de *Galanthus plicatus* subsp. *byzantinus*

escape (volverse silvestre) una planta que ha sobrepasado los límites del cultivo (p.e.: un jardín) y prospera en la naturaleza

ex *ex* después, puede utilizarse entre los nombres de dos autores, el segundo de los cuales publicó el nombre indicado o sugerido por el primero

excl. *exclusus* excluida

hort. *hortorum* de jardines (horticultura); cultivada o prospera en jardines; no se trata de una planta silvestre

ICNB (CINB) Código Internacional de Nomenclatura Botánica

incl. inclusive

in prep. en preparación

in sched. *in scheda* en un espécimen de herbario o etiqueta

in syn. *in synonymia* en sinonimia

ined. *ineditus* : inédito

introduction (introducción) una planta que ocurre en un país, o en cualquier otra localidad, debido a la influencia antropogénica (intencionalmente o al azar); cualquier planta que no es nativa

key (clave) un sistema escrito utilizado para la identificación de organismos (p.e.: plantas)

leg. *legit* el recolector; el coleccionista

misspelling (error de ortografía) un nombre que se ha escrito incorrectamente; no se trata de un nombre nuevo o diferente

morphology (morfología) la forma y estructura de un organismo (p.e.: una planta)

name causing confusion (nombre de dudosa semejanza) un nombre que no se usa, ya que no puede asignarse a un determinado taxón sin crear confusión (p.e.: una especie de planta)

native (nativo) un organismo (p.e.: una planta) que prospera naturalmente en un país o región, etc.

naturalized (naturalizada) una planta que ha sido introducida (véase introducción) o se ha vuelto silvestre (véase volverse silvestre) pero que parece una planta silvestre y se reproduce por sí misma en su nuevo medio

nom. *nomen* nombre

nom. ambig. *nomen ambiguum* nombre ambiguo

nom. cons. prop. *nomen conservandum propositum* nombre propuesto para la conservación con arreglo a lo dispuesto en el Código Internacional de Nomenclatura Botánica (ICBN)

nomenclature (nomenclatura) parte de la ciencia que se ocupa de atribuir nombres a organismos (p.e.: plantas)

non *non* no

only known from cultivation (sólo se conoce en cultivo) una planta que no ocurre en el medio silvestre, únicamente en cultivo

orthographic variant (variante ortográfica) una alternativa ortográfica del mismo nombre

pro parte *pro parte* : parcialmente, en parte

provisional name (nombre provisional) nombre asignado temporalmente hasta que se disponga de una descripción válida

sens. *sensu* en el sentido de; la forma en que un autor interpreta o utiliza un nombre

sens. lat. *sensu lato* en sentido generalizado, un taxón (normalmente una especie) y todos sus taxa subordinados (p.e.: subspecies) y/o otros taxa a veces considerados como distintos

sic *sic* utilizado después de una palabra que pudiera parecer inexacta o absurda, para dar a entender que es textual

synonym (sinónimo) un nombre que se aplica a un taxón pero que no puede utilizarse ya que no es un nombre aceptado – el sinónimo o los sinónimos forman la sinonimia

taxa plural de taxón

taxon (taxón) una determinada unidad de clasificación, p.e.: género, especie, subespecie

var. variedad

* Expresamos nuestro agradecimiento al Dr. Aaron Davis, Real Jardín Botánico de Kew, por la presentación de esta guía.

8. Bibliografía

Brummitt, R.K. and Powell, C.E. (Eds.) (1992). *Authors of Plant Names*. Royal Botanic Gardens, Kew (UK).

Newton, L.E. (2001). Sansevieria. In: U.Eggli (ed.), *Illustrated Handbook of Succulent Plants*: *Monocotyledons:* 261-272. Springer, Berlin.

Rapanarivo, S.H.J.V, Lavranos, J.J., Leeuwenberg, A.J.M. & Röösli, W. (1999). *Pachypodium (Apocynaceae), Taxonomy, habitats & cultivation*. A.A. Balkema, Rotterdam / Brookfield.

Rowley, G.D. (2002). Pachypodium (Apocynaceae). In U. Eggli (ed.), *Illustrated Handbook of Succulent Plants: Dicotyledons* (in press). Springer, Heidelberg.

Terminology Bulletin No. 347/Rev. 1. 1997, United Nations.

PART I: ALL NAMES IN CURRENT USE
Ordered alphabetically on all names for the genera:

Aloe and *Pachypodium*

PREMIERE PARTIE: TOUS LES NOMS D'USAGE COURANT
Par ordre alphabétique de tous le noms

Aloe et *Pachypodium*

PARTE I: TODOS LOS NOMBRES UTILIZADOS NORMALMENTE
Presentados por orden alfabético

Aloe y *Pachypodium*

Part I: All Names / Tous les Noms / Todos los Nombres

ALPHABETICAL LISTING OF ALL NAMES FOR THE GENERA:
Aloe and *Pachypodium*

LISTES ALPHABETIQUES DE TOUS LES NOMS POUR LES GENRES:
Aloe et *Pachypodium*

PRESENTACIÓN POR ORDEN ALFABETICO DE TODOS LOS NOMBRES
PARA LOS GENEROS:
Aloe y *Pachypodium*

ALL NAMES TOUS LES NOMS TODOS LOS NOMBRES	ACCEPTED NAME NOM RECONNU NOMBRE ACEPTADO
Adenium namaquanum	Pachypodium namaquanum
Aloe aageodonta	
Aloe abyssicola	
Aloe abyssinica var. *peacockii*	Aloe elegans
Aloe abyssinica var. *percrassa*	Aloe percrassa
Aloe abyssinica Hook.f.	Aloe adigratana
Aloe abyssinica Salm-Dyck	Aloe camperi
Aloe abyssinica A.Berger	Aloe elegans
Aloe aculeata	
Aloe acuminata var. *major*	Aloe humilis
Aloe acuminata	Aloe humilis
Aloe acutissima	
Aloe acutissima var. acutissima	
Aloe acutissima var. antanimorensis	
Aloe adigratana	
Aloe aethiopica	Aloe elegans
Aloe affinis	
Aloe africana	
Aloe africana var. *angustior*	Aloe africana
Aloe africana var. *latifolia*	Aloe africana
Aloe agrophila	Aloe boylei
Aloe ahmarensis	
Aloe albida	
Aloe albiflora	
Aloe albocincta	Aloe striata ssp. striata
Aloe albopicta	Aloe camperi
Aloe albovestita	
Aloe aldabrensis	
Aloe alfredii	
Aloe alooides	
Aloe amanensis	Aloe lateritia var. lateritia
Aloe ambigens	
Aloe amicorum	
Aloe ammophila	Aloe zebrina
Aloe amoena	Aloe framesii
Aloe amudatensis	
Aloe andongensis	
Aloe andongensis var. andongensis	
Aloe andongensis var. repens	
Aloe andringitrensis	
Aloe angelica	
Aloe angiensis	Aloe wollastonii

ALL NAMES	ACCEPTED NAMES
Aloe angiensis var. *kitaliensis*	Aloe wollastonii
Aloe angolensis	
Aloe angustifolia Haw.	Aloe africana
Aloe angustifolia Groenew.	Aloe zebrina
Aloe anivoranoensis	
Aloe ankaranensis	
Aloe ankoberensis	
Aloe antandroi	
Aloe antsingyensis	
Aloe arabica	Aloe pendens
Aloe arborea	Aloe arborescens
Aloe arborescens	
Aloe arborescens var. *frutescens*	Aloe arborescens
Aloe arborescens var. *milleri*	Aloe arborescens
Aloe arborescens var. *natalensis*	Aloe arborescens
Aloe arborescens var. *pachythyrsa*	Aloe arborescens
Aloe archeri	
Aloe arenicola	
Aloe argenticauda	
Aloe aristata	
Aloe aristata var. *leiophylla*	Aloe aristata
Aloe aristata var. *parvifolia*	Aloe aristata
Aloe armatissima	
Aloe asperifolia	
Aloe atherstonei	Aloe pluridens
Aloe audhalica	Aloe vacillans
Aloe aurantiaca	Aloe striatula var. striatula
Aloe ausana	Aloe variegata
Aloe babatiensis	
Aloe bainesii var. *barberae*	Aloe barberae
Aloe bainesii	Aloe barberae
Aloe bakeri	
Aloe ballii	
Aloe ballii var. ballii	
Aloe ballii var. makurupiniensis	
Aloe ballyi	
Aloe bamangwatensis	Aloe zebrina
Aloe barbadensis var. *chinensis*	Aloe vera
Aloe barbadensis	Aloe vera
Aloe barberae	
Aloe barbertoniae	
Aloe bargalensis	
Aloe barteri var. *dahomensis*	Aloe buettneri
Aloe barteri var. *lutea*	Aloe schweinfurthii
Aloe barteri var. *sudanica*	Aloe buettneri
Aloe barteri Baker	Aloe buettneri
Aloe barteri Schnell	Aloe macrocarpa
Aloe belavenokensis	
Aloe bella	
Aloe bellatula	
Aloe beniensis	Aloe dawei
Aloe bequaertii	Aloe wollastonii
Aloe berevoana	
Aloe berhana	Aloe debrana
Aloe bernadettae	
Aloe bertemariae	
Aloe betsileensis	
Aloe bicomitum	
Aloe boastii	Aloe chortolirioides var. chortolirioides

Part I: All Names / Tous les Noms / Todos los Nombres

ALL NAMES	ACCEPTED NAMES
Aloe boehmii	Aloe lateritia var. lateritia
Aloe boiteaui	
Aloe bolusii	Aloe africana
Aloe boranensis	Aloe otallensis
Aloe boscawenii	
Aloe bosseri	
Aloe bowiea	
Aloe boylei	
Aloe boylei ssp. boylei	
Aloe boylei ssp. major	
Aloe brachystachys	
Aloe branddraaiensis	
Aloe brandhamii	
Aloe brevifolia	
Aloe brevifolia var. brevifolia	
Aloe brevifolia var. depressa	
Aloe brevifolia var. postgenita	
Aloe brevifolia var. *serra*	Aloe brevifolia var. depressa
Aloe brevifolia	Aloe distans
Aloe breviscapa	
Aloe broomii	
Aloe broomii var. broomii	
Aloe broomii var. tarkaensis	
Aloe brunneodentata	
Aloe brunneo-punctata	Aloe nuttii
Aloe brunneostriata	
Aloe brunnthaleri	Aloe microstigma
Aloe buchananii	
Aloe buchlohii	
Aloe buettneri	
Aloe buhrii	
Aloe bukobana	
Aloe bulbicaulis	
Aloe bulbillifera	
Aloe bulbillifera var. bulbillifera	
Aloe bulbillifera var. paulianae	
Aloe bullockii	
Aloe burgersfortensis	
Aloe bussei	
Aloe calcairophila	
Aloe calcairophylla	Aloe calcairophila
Aloe calidophila	
Aloe cameronii	
Aloe cameronii var. bondana	
Aloe cameronii var. cameronii	
Aloe cameronii var. dedzana	
Aloe camperi	
Aloe campylosiphon	Aloe lateritia var. lateritia
Aloe canarina	
Aloe candelabrum A.Berger	Aloe ferox
Aloe candelabrum Engl. & Drude	Aloe thraskii
Aloe cannellii	
Aloe capitata	
Aloe capitata var. capitata	
Aloe capitata var. cipolinicola	
Aloe capitata var. gneissicola	
Aloe capitata var. quartziticola	
Aloe capitata var. silvicola	
Aloe capitata var. *trachyticola*	Aloe trachyticola

ALL NAMES	ACCEPTED NAMES
Aloe capmanambatoensis	
Aloe caricina	Aloe myriacantha
Aloe carnea	
Aloe carowii	Aloe sladeniana
Aloe cascadensis	Aloe striatula var. striatula
Aloe castanea	
Aloe castellorum	
Aloe catengiana	
Aloe cephalophora	
Aloe cernua	Aloe capitata var. capitata
Aloe chabaudii	
Aloe chabaudii var. chabaudii	
Aloe chabaudii var. mlanjeana	
Aloe chabaudii var. verekeri	
Aloe cheranganiensis	
Aloe chimanimaniensis	Aloe swynnertonii
Aloe chinensis	Aloe vera
Aloe chlorantha	
Aloe chortolirioides	
Aloe chortolirioides var. *boastii*	Aloe chortolirioides var. chortolirioides
Aloe chortolirioides var. chortolirioides	
Aloe chortolirioides var. woolliana	
Aloe christianii	
Aloe chrysostachys	
Aloe ciliaris	
Aloe ciliaris forma *flanaganii*	Aloe ciliaris var. ciliaris
Aloe ciliaris forma *gigas*	Aloe ciliaris var. ciliaris
Aloe ciliaris forma *tidmarshii*	Aloe ciliaris var. tidmarshii
Aloe ciliaris var. ciliaris	
Aloe ciliaris var. *flanaganii*	Aloe ciliaris var. ciliaris
Aloe ciliaris var. redacta	
Aloe ciliaris var. tidmarshii	
Aloe citrea	
Aloe citrina	
Aloe classenii	
Aloe claviflora	
Aloe collenetteae	
Aloe collina	
Aloe commixta	
Aloe commutata	Aloe macrocarpa
Aloe comosa	
Aloe comosibracteata	Aloe greatheadii var. davyana
Aloe compacta	Aloe macrosiphon
Aloe compressa	
Aloe compressa var. compressa	
Aloe compressa var. paucituberculata	
Aloe compressa var. rugosquamosa	
Aloe compressa var. schistophila	
Aloe comptonii	
Aloe concinna	Aloe squarrosa
Aloe confusa	
Aloe congdonii	
Aloe conifera	
Aloe contigua	Aloe imalotensis
Aloe cooperi	
Aloe cooperi ssp. cooperi	
Aloe cooperi ssp. pulchra	
Aloe corallina	
Aloe corbisieri	Aloe nuttii

23

ALL NAMES	ACCEPTED NAMES
Aloe crassipes	
Aloe cremersii	
Aloe cremnophila	
Aloe cryptoflora	
Aloe cryptopoda	
Aloe cyrtophylla	
Aloe dabenorisana	
Aloe davyana	Aloe greatheadii var. davyana
Aloe davyana var. *subolifera*	Aloe greatheadii var. davyana
Aloe dawei	
Aloe debrana	
Aloe decora	Aloe claviflora
Aloe decorsei	
Aloe decurva	
Aloe decurvidens	Aloe parvibracteata
Aloe delphinensis	
Aloe deltoideodonta	
Aloe deltoideodonta forma *latifolia*	Aloe imalotensis
Aloe deltoideodonta forma *longifolia*	Aloe imalotensis
Aloe deltoideodonta subforma *variegata*	Aloe imalotensis
Aloe deltoideodonta var. brevifolia	
Aloe deltoideodonta var. candicans	
Aloe deltoideodonta var. *contigua*	Aloe imalotensis
Aloe deltoideodonta var. deltoideodonta	
Aloe deltoideodonta var. *intermedia*	Aloe subacutissima
Aloe deltoideodonta var. *typica*	Aloe deltoideodonta var. deltoideodonta
Aloe dependens	Aloe pendens
Aloe depressa	Aloe brevifolia var. depressa
Aloe descoingsii	
Aloe descoingsii ssp. augustina	
Aloe descoingsii ssp. descoingsii	
Aloe deserti	
Aloe dewetii	
Aloe dewinteri	
Aloe dhalensis	Aloe vacillans
Aloe dhufarensis	
Aloe dichotoma	
Aloe dichotoma var. *montana*	Aloe dichotoma
Aloe dichotoma var. *ramosissima*	Aloe ramosissima
Aloe dinteri	
Aloe diolii	
Aloe distans	
Aloe disticha	Aloe maculata
Aloe disticha var. *plicatilis*	Aloe plicatilis
Aloe divaricata	
Aloe divaricata var. divaricata	
Aloe divaricata var. rosea	
Aloe doci	
Aloe doei var. doei	
Aloe doei var. lavranosii	
Aloe dolomitica	Aloe vryheidensis
Aloe dominella	
Aloe dorotheae	
Aloe duckeri	
Aloe dumetorum	Aloe ellenbeckii
Aloe dyeri	
Aloe echinata	Aloe humilis
Aloe ecklonis	
Aloe edentata	

ALL NAMES	ACCEPTED NAMES
Aloe edulis	Aloe macrocarpa
Aloe elata	
Aloe elegans	
Aloe elgonica	
Aloe ellenbeckii	
Aloe ellenbergeri	Aloe aristata
Aloe elongata	Aloe vera
Aloe eminens	
Aloe engleri	Aloe secundiflora var. secundiflora
Aloe enotata	
Aloe eremophila	
Aloe erensii	
Aloe ericetorum	
Aloe erinacea	Aloe melanacantha var. erinacea
Aloe eru	Aloe camperi
Aloe eru forma *erecta*	Aloe camperi
Aloe eru forma *glauca*	Aloe camperi
Aloe eru forma *maculata*	Aloe camperi
Aloe eru forma *parvipunctata*	Aloe camperi
Aloe eru var. *cornuta*	Aloe camperi
Aloe eru var. *hookeri*	Aloe adigratana
Aloe erythrophylla	
Aloe esculenta	
Aloe eumassawana	
Aloe excelsa	
Aloe excelsa var. breviflora	
Aloe excelsa var. excelsa	
Aloe eylesii	Aloe rhodesiana
Aloe falcata	
Aloe ferox Mill.	
Aloe ferox A.Berger	Aloe marlothii
Aloe ferox var. *erythrocarpa*	Aloe ferox
Aloe ferox var. *galpinii*	Aloe ferox
Aloe ferox var. *hanburyi*	Aloe ferox
Aloe ferox var. *incurva*	Aloe ferox
Aloe ferox var. *subferox*	Aloe ferox
Aloe ferox var. *xanthostachys*	Aloe marlothii
Aloe fibrosa	
Aloe fievetii	
Aloe fimbrialis	
Aloe flabelliformis	Aloe plicatilis
Aloe flava	Aloe vera
Aloe fleurentiniorum	
Aloe fleuretteana	
Aloe flexilifolia	
Aloe floramaculata	Aloe secundiflora var. secundiflora
Aloe forbesii	
Aloe fosteri	
Aloe fouriei	
Aloe fragilis	
Aloe framesii	
Aloe francombei	
Aloe friisii	
Aloe frutescens	Aloe arborescens
Aloe fruticosa	Aloe arborescens
Aloe fulleri	
Aloe galpinii	Aloe ferox
Aloe gariepensis	
Aloe gariusiana	Aloe gariepensis

ALL NAMES	ACCEPTED NAMES
Aloe gerstneri	
Aloe gilbertii	
Aloe gilbertii ssp. gilbertii	
Aloe gilbertii ssp. megalacanthoides	
Aloe gillettii	
Aloe gillilandii	Aloe sabaea
Aloe glabrescens	
Aloe glauca	
Aloe glauca var. *elatior*	Aloe glauca var. glauca
Aloe glauca var. glauca	
Aloe glauca var. *humilior*	Aloe glauca var. glauca
Aloe glauca var. *major*	Aloe glauca var. glauca
Aloe glauca var. *minor*	Aloe glauca var. glauca
Aloe glauca var. *muricata*	Aloe glauca var. spinosior
Aloe glauca var. spinosior	
Aloe globuligemma	
Aloe gloveri	Aloe hildebrandtii
Aloe gossweileri	
Aloe gracilicaulis	
Aloe graciliflora	Aloe greatheadii var. davyana
Aloe gracilis Baker	Aloe commixta
Aloe gracilis Haw.	
Aloe gracilis var. decumbens	
Aloe gracilis var. gracilis	
Aloe graminicola	Aloe lateritia var. graminicola
Aloe graminifolia	Aloe myriacantha
Aloe grandidentata	
Aloe grata	
Aloe greatheadii	
Aloe greatheadii var. davyana	
Aloe greatheadii var. greatheadii	
Aloe greenii	
Aloe greenwayi	Aloe leptosiphon
Aloe grisea	
Aloe guerrae	
Aloe guillaumetii	
Aloe haemanthifolia	
Aloe hanburyana	Aloe striata ssp. striata
Aloe hardyi	
Aloe harlana	
Aloe harmsii	Aloe dorotheae
Aloe haworthioides	
Aloe haworthioides var. aurantiaca	
Aloe haworthioides var. haworthioides	
Aloe hazeliana	
Aloe helenae	
Aloe heliderana	
Aloe hemmingii	
Aloe hendrickxii	
Aloe hereroensis	
Aloe hereroensis var. *hereroensis*	Aloe hereroensis
Aloe heybensis	
Aloe hijazensis	
Aloe hildebrandtii	
Aloe hlangapensis	Aloe hlangapies
Aloe hlangapies	
Aloe hlangapitis	Aloe hlangapies
Aloe howmanii	
Aloe humbertii	

ALL NAMES	ACCEPTED NAMES
Aloe humilis (L.) Mill.	
Aloe humilis Ker Gawl.	Aloe humilis
Aloe humilis subvar. *minor*	Aloe humilis
Aloe humilis subvar. *semiguttata*	Aloe humilis
Aloe humilis var. *acuminata*	Aloe humilis
Aloe humilis var. *candollei*	Aloe humilis
Aloe humilis var. *echinata*	Aloe humilis
Aloe humilis var. *humilis*	Aloe humilis
Aloe humilis var. *incurva*	Aloe humilis
Aloe humilis var. *suberecta*	Aloe humilis
Aloe humilis var. *subtuberculata*	Aloe humilis
Aloe ibitiensis	
Aloe ibityensis	Aloe ibitiensis
Aloe imalotensis	
Aloe ×imerinensis	
Aloe immaculata	
Aloe inamara	
Aloe inconspicua	
Aloe incurva	Aloe humilis
Aloe indica	Aloe vera
Aloe inermis	
Aloe integra	
Aloe intermedia	Aloe subacutissima
Aloe inyangensis	
Aloe inyangensis var. inyangensis	
Aloe inyangensis var. kimberleyana	
Aloe isaloensis	
Aloe itremensis	
Aloe jacksonii	
Aloe jex-blakeae	Aloe ruspoliana
Aloe johnstonii	Aloe myriacantha
Aloe jucunda	
Aloe juttae	Aloe microstigma
Aloe juvenna	
Aloe karasbergensis	Aloe striata ssp. karasbergensis
Aloe ×keayi	
Aloe kedongensis	
Aloe kefaensis	
Aloe keithii	
Aloe ketabrowniorum	
Aloe khamiesensis	
Aloe kilifiensis	
Aloe kirkii	Aloe massawana
Aloe kniphofioides	
Aloe komaggasensis	Aloe striata ssp. komaggasensis
Aloe komatiensis	Aloe parvibracteata
Aloe krapohliana	
Aloe krapohliana var. dumoulinii	
Aloe krapohliana var. krapohliana	
Aloe kraussii Baker	
Aloe kraussii Schönland	Aloe albida
Aloe kraussii var. *minor*	Aloe albida
Aloe kulalensis	
Aloe labiaflava	Aloe greatheadii var. davyana
Aloe labworana	
Aloe laeta	
Aloe laeta var. laeta	
Aloe laeta var. maniaensis	
Aloe lanuriensis	Aloe wollastonii

ALL NAMES	ACCEPTED NAMES
Aloe lanzae	Aloe vera
Aloe lastii	Aloe brachystachys
Aloe lateritia	
Aloe lateritia var. graminicola	
Aloe lateritia var. *kitaliensis*	Aloe wollastonii
Aloe lateritia var. lateritia	
Aloe latifolia	Aloe maculata
Aloe lavranosii	
Aloe laxiflora	Aloe gracilis var. gracilis
Aloe laxissima	Aloe zebrina
Aloe leachii	
Aloe leandrii	
Aloe leedalii	
Aloe lensayuensis	
Aloe lepida	
Aloe leptocaulon	Aloe antandroi
Aloe leptophylla	Aloe maculata
Aloe leptophylla var. *stenophylla*	Aloe maculata
Aloe leptosiphon	
Aloe lettyae	
Aloe lindenii	
Aloe linearifolia	
Aloe lineata	
Aloe lineata var. *glaucescens*	Aloe lineata var. lineata
Aloe lineata var. lineata	
Aloe lineata var. muirii	
Aloe lineata var. *viridis*	Aloe lineata var. lineata
Aloe lingua	Aloe plicatilis
Aloe linguaeformis	Aloe plicatilis
Aloe littoralis	
Aloe lolwensis	
Aloe lomatophylloides	
Aloe longiaristata	Aloe aristata
Aloe longibracteata	Aloe greatheadii var. davyana
Aloe longistyla	
Aloe luapulana	
Aloe lucile-allorgeae	
Aloe lugardiana	Aloe zebrina
Aloe luntii	
Aloe lusitanica	Aloe parvibracteata
Aloe lutescens	
Aloe maclaughlinii	Aloe mcloughlinii
Aloe macleayi	
Aloe macowanii	Aloe striatula var. striatula
Aloe macra	
Aloe macrocarpa	
Aloe macrocarpa var. *major*	Aloe macrocarpa
Aloc macroclada	
Aloe macrosiphon	
Aloe maculata All.	
Aloe maculata Lam.	Aloe maculata
Aloe maculata Forssk.	Aloe officinalis
Aloe maculosa	Aloe maculata
Aloe madecassa	
Aloe madecassa var. lutea	
Aloe madecassa var. madecassa	
Aloe magnidentata	Aloe megalacantha ssp. megalacantha
Aloe marginalis	Aloe purpurea
Aloe marginata	Aloe purpurea

ALL NAMES	ACCEPTED NAMES
Aloe marlothii A. Berger	
Aloe marlothii J.M.Wood	**Aloe marlothii**
Aloe marlothii ssp. **orientalis**	
Aloe marlothii var. **bicolor**	
Aloe marlothii var. **marlothii**	
Aloe marsabitensis	**Aloe secundiflora** var. **secundiflora**
Aloe marshallii	**Aloe kniphofioides**
Aloe massawana	
Aloe mawii	
Aloe mayottensis	
Aloe mccoyi	
Aloe mcloughlinii	
Aloe medishiana	
Aloe megalacantha	
Aloe megalacantha ssp. **alticola**	
Aloe megalacantha ssp. **megalacantha**	
Aloe megalocarpa	
Aloe melanacantha	
Aloe melanacantha var. **erinacea**	
Aloe melanacantha var. **melanacantha**	
Aloe melsetterensis	**Aloe swynnertonii**
Aloe menachensis	
Aloe mendesii	
Aloe menyharthii	
Aloe menyharthii ssp. **ensifolia**	
Aloe menyharthii ssp. **menyharthii**	
Aloe meruana	**Aloe chrysostachys**
Aloe metallica	
Aloe meyeri	
Aloe micracantha Baker	**Aloe boylei**
Aloe micracantha Haw.	
Aloe micracantha Link & Otto	**Aloe micracantha**
Aloe microdonta	
Aloe microstigma	
Aloe millotii	
Aloe milne-redheadii	
Aloe minima Baker	
Aloe minima (Reynolds) Reynolds	**Aloe saundersiae**
Aloe minima var. **blyderivierensis**	
Aloe minima var. **minima**	
Aloe mitriformis	
Aloe mitriformis var. *angustior*	**Aloe distans**
Aloe mitriformis var. *brevifolia*	**Aloe distans**
Aloe mitriformis var. *elatior*	**Aloe mitriformis**
Aloe mitriformis var. *humilior*	**Aloe mitriformis**
Aloe mketiensis	**Aloe nuttii**
Aloe modesta	
Aloe molederana	
Aloe monotropa	
Aloe montana	**Aloe dichotoma**
Aloe monticola	
Aloe morijensis	
Aloe morogoroensis	**Aloe bussei**
Aloe mubendiensis	
Aloe mudenensis	
Aloe muirii	**Aloe lineata** var. **muirii**
Aloe multicolor	
Aloe munchii	
Aloe muricata Haw.	**Aloe ferox**

ALL NAMES	ACCEPTED NAMES
Aloe muricata Schult.	**Aloe glauca** var. **spinosior**
Aloe murina	
Aloe musapana	
Aloe mutabilis	
Aloe mutans	**Aloe greatheadii** var. **davyana**
Aloe mwanzana	**Aloe macrosiphon**
Aloe myriacantha	
Aloe myriacantha var. *minor*	**Aloe albida**
Aloe mzimbana	
Aloe namibensis	
Aloe namorokaensis	
Aloe natalensis	**Aloe arborescens**
Aloe ngobitensis	**Aloe nyeriensis**
Aloe ngongensis	
Aloe niebuhriana	
Aloe nitens	**Aloe rupestris**
Aloe nubigena	
Aloe nuttii	
Aloe nyeriensis	
Aloe nyeriensis ssp. *kedongensis*	**Aloe kedongensis**
Aloe occidentalis	
Aloe officinalis	
Aloe officinalis var. *angustifolia*	**Aloe officinalis**
Aloe oligophylla	
Aloe oligospila	**Aloe percrassa**
Aloe orientalis	
Aloe ortholopha	
Aloe otallensis	
Aloe otallensis var. *elongata*	**Aloe rugosifolia**
Aloe pachygaster	
Aloe paedogona	
Aloe pallidiflora	**Aloe greatheadii** var. **greatheadii**
Aloe palmiformis	
Aloe paludicola	**Aloe buettneri**
Aloe paniculata	**Aloe striata** ssp. **striata**
Aloe parallelifolia	
Aloe parvibracteata	
Aloe parvibracteata var. *zuluensis*	**Aloe parvibracteata**
Aloe parvicapsula	
Aloe parvicoma	
Aloe parvidens	
Aloe parviflora	**Aloe minima** var. **minima**
Aloe parvispina	**Aloe mitriformis**
Aloe parvula A.Berger	
Aloe parvula Reynolds	**Aloe perrieri**
Aloe patersonii	
Aloe peacockii	**Aloe elegans**
Aloe pearsonii	
Aloe peckii	
Aloe peglerae	
Aloe pembana	
Aloe pendens	
Aloe penduliflora	
Aloe percrassa Tod.	
Aloe percrassa Schweinf.	**Aloe trichosantha** ssp. **trichosantha**
Aloe percrassa var. *albo-picta*	**Aloe trichosantha** ssp. **trichosantha**
Aloe percrassa var. *menachensis*	**Aloe menachensis**
Aloe percrassa var. *saganeitiana*	**Aloe elegans**
Aloe perfoliata	**Aloe ferox**

ALL NAMES	ACCEPTED NAMES
Aloe perfoliata var. α L.	Aloe commixta
Aloe perfoliata var. β L.f.	Aloe africana
Aloe perfoliata var. γ L.	Aloe ferox
Aloe perfoliata var. δ L.	Aloe maculata
Aloe perfoliata var. ε L.	Aloe ferox
Aloe perfoliata var. ζ L.	Aloe brevifolia var. depressa
Aloe perfoliata var. ζ Willd.	Aloe ferox
Aloe perfoliata var. η L.	Aloe arborescens
Aloe perfoliata var. θ L.	Aloe maculata
Aloe perfoliata var. κ L.	Aloe glauca var. glauca
Aloe perfoliata var. λ L.	Aloe maculata
Aloe perfoliata var. ν L.	Aloe mitriformis
Aloe perfoliata var. ξ L.	Aloe succotrina
Aloe perfoliata var. ξ Willd.	Aloe mitriformis
Aloe perfoliata var. *africana*	Aloe africana
Aloe perfoliata var. *arborescens*	Aloe arborescens
Aloe perfoliata var. *barbadensis*	Aloe vera
Aloe perfoliata var. *brevifolia*	Aloe distans
Aloe perfoliata var. *ferox*	Aloe ferox
Aloe perfoliata var. *glauca*	Aloe glauca var. glauca
Aloe perfoliata var. *humilis*	Aloe humilis
Aloe perfoliata var. *lineata*	Aloe lineata
Aloe perfoliata var. *mitriformis*	Aloe mitriformis
Aloe perfoliata var. *purpurascens*	Aloe succotrina
Aloe perfoliata var. *saponaria*	Aloe maculata
Aloe perfoliata var. *succotrina*	Aloe succotrina
Aloe perfoliata var. *vera*	Aloe vera
Aloe perrieri	
Aloe perryi	
Aloe petricola	
Aloe petrophila	
Aloe peyrierasii	
Aloe pictifolia	
Aloe pienaarii	Aloe cryptopoda
Aloe pillansii	
Aloe pirottae	
Aloe platyphylla	Aloe zebrina
Aloe plicatilis	
Aloe plicatilis var. *major*	Aloe plicatilis
Aloe plowesii	
Aloe pluridens	
Aloe pluridens var. *beckeri*	Aloe pluridens
Aloe pole-evansii	Aloe dawei
Aloe polyphylla	
Aloe pongolensis	Aloe parvibracteata
Aloe pongolensis var. *zuluensis*	Aloe parvibracteata
Aloe porphyrostachys	
Aloe postgenita	Aloe brevifolia var. postgenita
Aloe powysiorum	
Aloe pratensis	
Aloe pretoriensis	
Aloe prinslooi	
Aloe procera	
Aloe prolifera	Aloe brevifolia var. brevifolia
Aloe prolifera var. *major*	Aloe brevifolia var. postgenita
Aloe propagulifera	
Aloe prostrata	
Aloe prostrata ssp. pallida	

ALL NAMES	ACCEPTED NAMES
Aloe pruinosa	
Aloe pseudoafricana	Aloe africana
Aloe pseudoferox	Aloe ferox
Aloe pseudorubroviolacea	
Aloe pubescens	
Aloe pulcherrima	
Aloe pulchra	Aloe bella
Aloe punctata	Aloe variegata
Aloe purpurascens	Aloe succotrina
Aloe purpurea	
Aloe pustuligemma	
Aloe pycnacantha	Aloe rupestris
Aloe ×qaharensis	
Aloe rabaiensis	
Aloe ramosa	Aloe dichotoma
Aloe ramosissima	
Aloe rauhii	
Aloe recurvifolia	Aloe alooides
Aloe reitzii	
Aloe reitzii var. reitzii	
Aloe reitzii var. vernalis	
Aloe retrospiciens	
Aloe reynoldsii	
Aloe rhodacantha	Aloe glauca var. glauca
Aloe rhodesiana	
Aloe rhodocincta	Aloe striata ssp. striata
Aloe richardsiae	
Aloe richtersveldensis	Aloe meyeri
Aloe rigens	
Aloe rigens var. *glabrescens*	Aloe glabrescens
Aloe rigens var. mortimeri	
Aloe rigens var. rigens	
Aloe rivae	
Aloe rivierei	
Aloe rosea	
Aloe rubrolutea	Aloe littoralis
Aloe rubroviolacea	
Aloe ruffingiana	
Aloe rufocincta	Aloe purpurea
Aloe rugosifolia	
Aloe rupestris	
Aloe rupicola	
Aloe ruspoliana	
Aloe ruspoliana var. *dracaeniformis*	Aloe retrospiciens
Aloe sabaea	
Aloe sahundra	Aloe divaricata var. divaricata
Aloe saponaria	Aloe maculata
Aloe saponaria var. *brachyphylla*	Aloe maculata
Aloe saponaria var. *ficksburgensis*	Aloe maculata
Aloe saponaria var. *latifolia*	Aloe maculata
Aloe saponaria var. *saponaria*	Aloe maculata
Aloe saundersiae	
Aloe scabrifolia	
Aloe schelpei	
Aloe schilliana	
Aloe schimperi	Aloe percrassa
Aloe schinzii	Aloe littoralis
Aloe schlechteri	Aloe claviflora
Aloe schliebenii	Aloe brachystachys

ALL NAMES	ACCEPTED NAMES
Aloe schmidtiana	**Aloe cooperi** ssp. **cooperi**
Aloe schoelleri	
Aloe schomeri	
Aloe schweinfurthii Tod.	**Aloe elegans**
Aloe schweinfurthii Baker	
Aloe schweinfurthii var. *labworana*	**Aloe labworana**
Aloe scobinifolia	
Aloe scorpioides	
Aloe secundiflora	
Aloe secundiflora var. **secundiflora**	
Aloe secundiflora var. **sobolifera**	
Aloe sempervivoides	**Aloe parvula**
Aloe seretii	
Aloe serra	**Aloe brevifolia** var. **depressa**
Aloe serriyensis	
Aloe sessiliflora	**Aloe spicata**
Aloe shadensis	
Aloe sheilae	
Aloe silicicola	
Aloe simii	
Aloe sinana	
Aloe sinkatana	
Aloe sinuata	**Aloe succotrina**
Aloe sladeniana	
Aloe soccotorina	**Aloe succotrina**
Aloe soccotrina var. *purpurascens*	**Aloe succotrina**
Aloe socialis	
Aloe socotorina	**Aloe ferox**
Aloe solaiana	**Aloe lateritia** var. **graminicola**
Aloe somaliensis	
Aloe somaliensis var. **marmorata**	
Aloe somaliensis var. **somaliensis**	
Aloe soutpansbergensis	
Aloe speciosa	
Aloe spectabilis	**Aloe marlothii** var. **marlothii**
Aloe spicata Schweinf.	**Aloe camperi**
Aloe spicata L.f.	
Aloe splendens	
Aloe squarrosa	
Aloe steffanieana	
Aloe stephaninii	**Aloe ruspoliana**
Aloe steudneri	
Aloe striata	
Aloe striata ssp. **karasbergensis**	
Aloe striata ssp. **komaggasensis**	
Aloe striata ssp. **striata**	
Aloe striata var. *oligospila*	**Aloe striata** ssp. **striata**
Aloe striatula	
Aloe striatula forma *conimbricensis*	**Aloe striatula** var. **caesia**
Aloe striatula forma *haworthii*	**Aloe striatula** var. **caesia**
Aloe striatula forma *typica*	**Aloe striatula** var. **caesia**
Aloe striatula var. **caesia**	
Aloe striatula var. **striatula**	
Aloe stuhlmannii	**Aloe volkensii** ssp. **volkensii**
Aloe suarezensis	
Aloe subacutissima	
Aloe suberecta	**Aloe humilis**
Aloe suberecta var. *acuminata*	**Aloe humilis**
Aloe subferox	**Aloe ferox**

ALL NAMES	ACCEPTED NAMES
Aloe subfulta	**Aloe suffulta**
Aloe subtuberculata	**Aloe humilis**
Aloe succotrina All.	
Aloe succotrina Lam.	**Aloe succotrina**
Aloe succotrina var. *saxigena*	**Aloe succotrina**
Aloe suffulta	
Aloe suprafoliata Pole-Evans	
Aloe suprafoliolata hort.	**Aloe suprafoliata**
Aloe supralaevis	**Aloe ferox**
Aloe supralaevis var. *hanburyi*	**Aloe marlothii** var. **marlothii**
Aloe suzannae	
Aloe swynnertonii	
Aloe tauri	**Aloe spicata**
Aloe tenuior	
Aloe tenuior var. *decidua*	**Aloe tenuior**
Aloe tenuior var. *densiflora*	**Aloe tenuior**
Aloe tenuior var. *glaucescens*	**Aloe tenuior**
Aloe tenuior var. *rubriflora*	**Aloe tenuior**
Aloe termetophila	**Aloe greatheadii** var. **greatheadii**
Aloe tewoldei	
Aloe thompsoniae	
Aloe thorncroftii	
Aloe thraskii	
Aloe tidmarshii	**Aloe ciliaris** var. **tidmarshii**
Aloe tomentosa	
Aloe tomentosa forma *viridiflora*	**Aloe tomentosa**
Aloe tormentorii	
Aloe tororoana	
Aloe torrei	
Aloe torrei var. *wildii*	**Aloe wildii**
Aloe trachyticola	
Aloe transvaalensis	**Aloe zebrina**
Aloe transvaalensis var. *stenacantha*	**Aloe zebrina**
Aloe trichosantha	
Aloe trichosantha ssp. **longiflora**	
Aloe trichosantha ssp. **trichosantha**	
Aloe trichosantha var. *menachensis*	**Aloe menachensis**
Aloe trigonantha	
Aloe tripetala	**Aloe plicatilis**
Aloe trivialis	**Aloe schweinfurthii**
Aloe trothae	**Aloe bulbicaulis**
Aloe tuberculata	**Aloe humilis**
Aloe tugenensis	
Aloe turkanensis	
Aloe tweediae	
Aloe ukambensis	
Aloe umbellata	**Aloe maculata**
Aloe umfoloziensis	
Aloe vacillans	
Aloe vahontsohy	**Aloe divaricata** var. **divaricata**
Aloe vallaris	
Aloe vanbalenii	
Aloe vandermerwei	
Aloe vaombe	
Aloe vaombe var. **poissonii**	
Aloe vaombe var. **vaombe**	
Aloe vaotsanda	
Aloe vaotsohy	**Aloe divaricata** var. **divaricata**
Aloe vaotsohy var. *rosea*	**Aloe divaricata** var. **rosea**

ALL NAMES	ACCEPTED NAMES
Aloe variegata	Aloe pendens
Aloe variegata	
Aloe variegata var. *haworthii*	Aloe variegata
Aloe venusta	Aloe bicomitum
Aloe vera Mill.	Aloe succotrina
Aloe vera (L.) Burm.f.	
Aloe vera var. *aethiopica*	Aloe elegans
Aloe vera var. *angustifolia*	Aloe officinalis
Aloe vera var. *chinensis*	Aloe vera
Aloe vera var. *lanzae*	Aloe vera
Aloe vera var. *littoralis*	Aloe vera
Aloe vera var. *officinalis*	Aloe officinalis
Aloe vera var. *wratislaviensis*	Aloe vera
Aloe verdoorniae	Aloe greatheadii var. davyana
Aloe verecunda	
Aloe versicolor	
Aloe veseyi	
Aloe viguieri	
Aloe viridiflora	
Aloe vituensis	
Aloe vivipara	Agave vivipara
Aloe vogtsii	
Aloe volkensii	
Aloe volkensii ssp. **multicaulis**	
Aloe volkensii ssp. **volkensii**	
Aloe vossii	
Aloe vryheidensis	
Aloe vulgaris	Aloe vera
Aloe whitcombei	
Aloe wickensii	Aloe cryptopoda
Aloe wickensii var. *lutea*	Aloe cryptopoda
Aloe wickensii var. *wickensii*	Aloe cryptopoda
Aloe wildii	
Aloe wilsonii	
Aloe wollastonii	
Aloe woodii	
Aloe woolliana	Aloe chortolirioides var. woolliana
Aloe wrefordii	
Aloe xanthacantha	Aloe mitriformis
Aloe yavellana	
Aloe yemenica	
Aloe zanzibarica	Aloe squarrosa
Aloe zebrina	
Aloe zombitsiensis	
Aloinella haworthioides	Aloe haworthioides
Belonites bispinosa	Pachypodium bispinosum
Belonites succulenta	Pachypodium succulentum
Bowiea africana	Aloe bowiea
Bowiea myriacantha	Aloe myriacantha
Catevala arborescens	Aloe arborescens
Catevala humilis	Aloe humilis
Chamaealoe africana	Aloe bowiea
Dracaena dentata	Aloe purpurea
Dracaena marginata	Aloe purpurea
Echites bispinosa	Pachypodium bispinosum
Echites succulenta	Pachypodium succulentum
Gasteria antandroi	Aloe antandroi
Guillauminia albiflora	Aloe albiflora
Guillauminia bakeri	Aloe bakeri

Part I: All Names / Tous les Noms / Todos los Nombres

ALL NAMES	ACCEPTED NAMES
Guillauminia bellatula	Aloe bellatula
Guillauminia calcairophila	Aloe calcairophila
Guillauminia descoingsii	Aloe descoingsii
Guillauminia rauhii	Aloe rauhii
Kumara disticha	Aloe plicatilis
Lemeea boiteaui	Aloe boiteaui
Lemeea haworthioides	Aloe haworthioides
Lemeea parvula	Aloe parvula
Leptaloe albida	Aloe albida
Leptaloe blyderivierensis	Aloe minima var. blyderivierensis
Leptaloe minima	Aloe minima
Leptaloe myriacantha	Aloe myriacantha
Leptaloe saundersiae	Aloe saundersiae
Lomatophyllum aldabrense	Aloe aldabrensis
Lomatophyllum aloiflorum	Aloe purpurea
Lomatophyllum anivoranoense	Aloe anivoranoensis
Lomatophyllum antsingyense	Aloe antsingyensis
Lomatophyllum belavenokense	Aloe belavenokensis
Lomatophyllum borbonicum	Aloe purpurea
Lomatophyllum citreum	Aloe citrea
Lomatophyllum lomatophylloides	Aloe lomatophylloides
Lomatophyllum macrum	Aloe macra
Lomatophyllum marginatum	Aloe purpurea
Lomatophyllum namorokaense	Aloe namorokaensis
Lomatophyllum occidentale	Aloe occidentalis
Lomatophyllum oligophyllum	Aloe oligophylla
Lomatophyllum orientale	Aloe orientalis
Lomatophyllum pembanum	Aloe pembana
Lomatophyllum peyrierasii	Aloe peyrierasii
Lomatophyllum propaguliferum	Aloe propagulifera
Lomatophyllum prostratum	Aloe prostrata
Lomatophyllum purpureum	Aloe purpurea
Lomatophyllum roseum	Aloe rosea
Lomatophyllum rufocinctum	Aloe purpurea
Lomatophyllum sociale	Aloe socialis
Lomatophyllum tormentorii	Aloe tomentorii
Lomatophyllum viviparum	Aloe schilliana
Notosceptrum alooides	Aloe alooides
Pachidendron africanum	Aloe africana
Pachidendron africanum var. *angustum*	Aloe africana
Pachidendron africanum var. *latum*	Aloe africana
Pachidendron angustifolium	Aloe africana
Pachidendron ferox	Aloe ferox
Pachidendron pseudoferox	Aloe ferox
Pachidendron supralaeve	Aloe ferox
Pachypodium ambongense	
Pachypodium baronii	
Pachypodium baronii var. baronii	
Pachypodium baronii var.*erythreum*	**Pachypodium baronii var. baronii**
Pachypodium baronii var.*typicum*	**Pachypodium baronii** var. baronii
Pachypodium baronii var. windsorii	
Pachypodium bicolor	**Pachypodium rosulatum** forma bicolor
Pachypodium bispinosum	
Pachypodium brevicalyx	**Pachypodium densiflorum**
Pachypodium brevicaule	
Pachypodium cactipes	**Pachypodium rosulatum**
Pachypodium champenoisianum	**Pachypodium lamerei**
Pachypodium decaryi	
Pachypodium densiflorum	

ALL NAMES	ACCEPTED NAMES
Pachypodium densiflorum var. *brevicalyx*	Pachypodium densiflorum
Pachypodium densiflorum var. densiflorum	
Pachypodium drakei	Pachypodium rosulatum
Pachypodium eburneum	Pachypodium rosulatum var. eburneum
Pachypodium geayi	
Pachypodium giganteum	Pachypodium lealii
Pachypodium glabrum	Pachypodium bispinosum
Pachypodium gracilius	Pachypodium rosulatum var. gracilius
Pachypodium griquense	Pachypodium succulentum
Pachypodium ×hojnyi	
Pachypodium horombense	
Pachypodium inopinatum	Pachypodium rosulatum var. inopinatum
Pachypodium jasminiflorum	Pachypodium succulentum
Pachypodium lamerei	
Pachypodium lamerei var. *lamerei*	Pachypodium lamerei
Pachypodium lamerei var. *ramosum*	Pachypodium lamerei
Pachypodium lamerei var. *typicum*	Pachypodium lamerei
Pachypodium lealii	
Pachypodium lealii ssp. lealii	
Pachypodium lealii ssp. saundersii	
Pachypodium menabeum	Pachypodium lamerei
Pachypodium meridionale	Pachypodium rutenbergianum var. meridionale
Pachypodium namaquanum	
Pachypodium obesum	Adenium obesum
Pachypodium ramosum	Pachypodium lamerei
Pachypodium ×rauhii	
Pachypodium rosulatum	
Pachypodium rosulatum forma bicolor	
Pachypodium rosulatum var. *delphinense*	Pachypodium rosulatum var. rosulatum
Pachypodium rosulatum var. *drakei*	Pachypodium rosulatum var. rosulatum
Pachypodium rosulatum var. eburneum	
Pachypodium rosulatum var. gracilius	
Pachypodium rosulatum var. *horombense*	Pachypodium horombense
Pachypodium rosulatum var. inopinatum	
Pachypodium rosulatum var. rosulatum	
Pachypodium rosulatum var. rosulatum forma bicolor	
Pachypodium rosulatum var. *stenanthum*	Pachypodium rosulatum var. rosulatum
Pachypodium rosulatum var. *typicum*	Pachypodium rosulatum var. rosulatum
Pachypodium rutenbergianum	
Pachypodium rutenbergianum forma *lamerei*	Pachypodium lamerei
Pachypodium rutenbergianum var. meridionale	
Pachypodium rutenbergianum var. *perrieri*	Pachypodium rutenbergianum var. sofiense
Pachypodium rutenbergianum var. rutenbergianum	
Pachypodium rutenbergianum var. sofiense	
Pachypodium rutenbergianum var. *typicum*	Pachypodium rutenbergianum var. rutenbergianum
Pachypodium saundersii	Pachypodium lealii ssp. saundersii
Pachypodium sofiense	Pachypodium rutenbergianum var. sofiense
Pachypodium succulentum	
Pachypodium tomentosum	Pachypodium succulentum
Pachypodium tuberosum	Pachypodium succulentum
Pachypodium tuberosum var. *loddigesii*	Pachypodium bispinosum
Pachypodium windsorii	Pachypodium baronii var. windsorii
Phylloma aloiflorum	Aloe purpurea
Phylloma macrum	Aloe macra

ALL NAMES	ACCEPTED NAMES
Phylloma rufocinctum	**Aloe purpurea**
Rhipidodendrum dichotomum	**Aloe dichotoma**
Rhipidodendrum distichum	**Aloe plicatilis**
Rhipidodendrum plicatile	**Aloe plicatilis**
Urginea alooides	**Aloe alooides**

PART II: ACCEPTED NAMES IN CURRENT USE
Ordered alphabetically on accepted names and including geographical distribution

Aloe and *Pachypodium*

DEUXIEME PARTIES: NOMS ACCEPTES D'USAGE COURANT
Par ordre alphabétique des noms acceptés

Aloe et *Pachypodium*

PARTE II: NOMBRES ACEPTADOS UTILIZADOS NORMALMENTE
Presentados por orden alfabético: nombres aceptados

Aloe y *Pachypodium*

Part II: Aloe

ALOE BINOMIALS IN CURRENT USE

ALOE BINOMES ACTUELLEMENT EN USAGE

ALOE BINOMIALES UTILIZADOS NORMALMENTE

Aloe aageodonta L.E.Newton 1993

Distribution: Kenya.

Aloe abyssicola Lavranos & Bilaidi 1971

Distribution: Yemen.

Aloe aculeata Pole-Evans 1915

Distribution: South Africa (Northern Prov.), Zimbabwe.

Aloe acutissima H.Perrier 1926

Distribution: Madagascar.

Aloe acutissima var. **acutissima**

Distribution: Madagascar.

Aloe acutissima var. **antanimorensis** Reynolds 1956

Distribution: Madagascar.

Aloe adigratana Reynolds 1957
Aloe abyssinica Hook.f. 1900 (nom. illeg., Art. 53.1)
Aloe eru var. *hookeri* A.Berger 1908

Distribution: Ethiopia.

Aloe affinis A.Berger 1908

Distribution: South Africa (Mpumalanga).

Aloe africana Mill. 1768
Aloe africana var. *angustior* Haw. 1819
Aloe africana var. *latifolia* Haw. 1819
Aloe angustifolia Haw. 1819
Aloe bolusii Baker 1880
Aloe perfoliata var. β L.f. 1753
Aloe perfoliata var. *africana* Aiton 1789
Aloe pseudoafricana Salm-Dyck 1817
Pachidendron africanum (Mill.) Haw. 1821

Pachidendron africanum var. *angustum* Haw. 1821
Pachidendron africanum var. *latum* Haw. 1821
Pachidendron angustifolium (Haw.) Haw. 1821

Distribution: South Africa (Eastern Cape).

Aloe ahmarensis Favell, M.B.Mill. & Al-Gifri 1999

Distribution: Yemen

Aloe albida (Stapf) Reynolds 1947
Aloe kraussii Schönland 1903 (nom. illeg., Art. 53.1)
Aloe kraussii var. *minor* Baker 1896
Aloe myriacantha var. *minor* (Baker) A.Berger 1908
Leptaloe albida Stapf 1933

Distribution: South Africa (Mpumalanga).

Aloe albiflora Guillaumin 1940
Guillauminia albiflora (Guillaumin) A.Bertrand 1956

Distribution: Madagascar.

Aloe albovestita S.Carter & Brandham 1983

Distribution: Somalia.

Aloe aldabrensis (Marais) L.E.Newton & G.D.Rowley 1997
Lomatophyllum aldabrense Marais 1975

Distribution: Seychelles (Aldabra Archipelago: Astove Atoll)

Aloe alfredii Rauh 1990

Distribution: Madagascar (C).

Aloe alooides (Bolus) Druten 1956
Aloe recurvifolia Groenew. 1935
Notosceptrum alooides (Bolus) Benth.
Urginea alooides Bolus 1881

Distribution: South Africa (Mpumalanga).

Aloe ambigens Chiov. 1928

Distribution: Somalia.

Aloe amicorum L.E.Newton 1991

Distribution: Kenya.

Part II: Aloe

Aloe amudatensis Reynolds 1956

Distribution: Kenya, Uganda.

Aloe andongensis Baker 1878

Distribution: Angola.

Aloe andongensis var. **andongensis**

Distribution: Angola.

Aloe andongensis var. **repens** L.C.Leach 1974

Distribution: Angola.

Aloe andringitrensis H.Perrier 1926

Distribution: Madagascar.

Aloe angelica Pole-Evans 1934

Distribution: South Africa (Northern Prov.).

Aloe angolensis Baker 1878

Distribution: Angola.

Aloe anivoranoensis (Rauh & Hebding) L.E.Newton & G.D.Rowley 1998
Lomatophyllum anivoranoense Rauh & Hebding 1998

Distribution: Madagascar (NE).

Aloe ankaranensis Rauh & Mangelsdorff 2000

Distribution: Madagascar (NW)

Aloe ankoberensis M.G.Gilbert & Sebsebe 1997

Distribution: Ethiopia.

Aloe antandroi (Decary) H.Perrier 1926
Aloe leptocaulon Bojer 1837 (nom. nud., Art. 32.1c)
Gasteria antandroi Decary 1921

Distribution: Madagascar.

Aloe antsingyensis (Léandri) L.E.Newton & G.D.Rowley 1997
Lomatophyllum antsingyense Léandri 1935

Distribution: Madagascar.

Aloe arborescens Mill. 1768
Aloe arborea Medik. 1783 (nom. illeg., Art. 53.1)
Aloe arborescens var. *frutescens* (Salm-Dyck) Link 1821
Aloe arborescens var. *milleri* A.Berger 1908
Aloe arborescens var. *natalensis* (J.M.Wood & M.S.Evans) A.Berger 1908
Aloe arborescens var. *pachythyrsa* A.Berger 1908
Aloe frutescens Salm-Dyck 1817
Aloe fruticosa Lam. 1783
Aloe natalensis J.M.Wood & M.S.Evans 1901
Aloe perfoliata var. η L. 1753
Aloe perfoliata var. *arborescens* Aiton 1789
Catevala arborescens Medik. 1789

Distribution: Malawi, Mozambique, South Africa, Zimbabwe.

Aloe archeri Lavranos 1977

Distribution: Kenya.

Aloe arenicola Reynolds 1938

Distribution: South Africa (Northern Cape, Western Cape).

Aloe argenticauda Merxm. & Giess 1974

Distribution: Namibia.

Aloe aristata Haw. 1825
Aloe longiaristata Schult. & Schult.f. 1829
Aloe aristata var. *leiophylla* Baker 1880
Aloe aristata var. *parvifolia* Baker 1896
Aloe ellenbergeri Guillaumin 1934

Distribution: Lesotho, South Africa (Northern Cape, Western Cape, Eastern Cape, Free State, KwaZulu-Natal).

Aloe armatissima Lavranos & Collen. 2000

Distribution: Saudi Arabia

Aloe asperifolia A.Berger 1905

Distribution: Namibia.

Part II: Aloe

Aloe babatiensis Christian & I.Verd. 1954

Distribution: Tanzania (United Republic of).

Aloe bakeri Scott-Elliot 1891
 Guillauminia bakeri (Scott-Elliot) P.V.Heath 1994

Distribution: Madagascar.

Aloe ballii Reynolds 1964

Distribution: Mozambique, Zimbabwe.

Aloe ballii var. **ballii**

Distribution: Mozambique, Zimbabwe.

Aloe ballii var. **makurupiniensis** Ellert 1998

Distribution: Mozambique, Zimbabwe.

Aloe ballyi Reynolds 1953

Distribution: Kenya, Tanzania (United Republic of).

Aloe barberae Dyer 1874
 Aloe bainesii Dyer 1874
 Aloe bainesii var. *barberae* (Dyer) Baker 1896

Distribution: Mozambique, South Africa (KwaZulu-Natal), Swaziland.

Aloe barbertoniae Pole-Evans 1917

Distribution: South Africa (Mpumalanga).

Aloe bargalensis Lavranos 1973

Distribution: Somalia.

Aloe belavenokensis (Rauh & Gerold) L.E.Newton & G.D.Rowley 1997
 Lomatophyllum belavenokense Rauh & Gerold 1994

Distribution: Madagascar.

Aloe bella G.D.Rowley 1974
 Aloe pulchra Lavranos 1973 (nom. illeg., Art. 53.1)

Distribution: Somalia.

Aloe bellatula Reynolds 1956
Guillauminia bellatula (Reynolds) P.V.Heath 1994

Distribution: Madagascar.

Aloe berevoana Lavranos 1998

Distribution: Madagascar (W).

Aloe bernadettae J.-B.Castillon 2000

Distribution: Madagascar

Aloe bertemariae Sebsebe & Dioli

Distribution: Ethiopia

Aloe betsileensis H.Perrier 1926

Distribution: Madagascar.

Aloe bicomitum L.C.Leach 1977
Aloe venusta Reynolds 1959 (nom. illeg., Art. 53.1)

Distribution: Tanzania (United Republic of), Zambia.

Aloe boiteaui Guillaumin 1942
Lemeea boiteaui (Guillaumin) P.V.Heath 1994

Distribution: Madagascar (S).

Aloe boscawenii Christian 1942

Distribution: Tanzania (United Republic of).

Aloe bosseri J.-B.Castillon 2000

Distribution: Madagascar

Aloe bowiea Schult. & Schult.f. 1829
Bowiea africana Haw. 1824
Chamaealoe africana (Haw.) A.Berger 1905

Distribution: South Africa (Eastern Cape).

Aloe boylei Baker 1892
Aloe agrophila Reynolds 1936

Part II: Aloe

Aloe micracantha Pole-Evans 1923 (nom. illeg., Art. 53.1)

Distribution: South Africa.

Aloe boylei ssp. **boylei**

Distribution: South Africa (Eastern Cape, KwaZulu-Natal, Mpumalanga, Northern Prov.).

Aloe boylei ssp. **major** Hilliard & B.L.Burtt 1985

Distribution: South Africa (KwaZulu-Natal).

Aloe brachystachys Baker 1895
Aloe lastii Baker 1901
Aloe schliebenii Lavranos 1970

Distribution: Tanzania (United Republic of).

Aloe branddraaiensis Groenew. 1940

Distribution: South Africa (Mpumalanga).

Aloe brandhamii S.Carter 1994

Distribution: Tanzania (United Republic of).

Aloe brevifolia Mill. 1771

Distribution: South Africa.

Aloe brevifolia var. **brevifolia**
Aloe prolifera Haw. 1804

Distribution: South Africa (Western Cape).

Aloe brevifolia var. **depressa** (Haw.) Baker 1880
Aloe brevifolia var. *serra* (DC.) A.Berger 1908
Aloe depressa Haw. 1804
Aloe perfoliata var. ζ L. 1753
Aloe serra DC. 1799

Distribution: South Africa (Western Cape).

Aloe brevifolia var. **postgenita** (M.Roem. & Schult.) Baker 1880
Aloe postgenita M.Roem. & Schult. 1830
Aloe prolifera var. *major* Salm-Dyck 1817

Distribution: South Africa (Western Cape).

Aloe breviscapa Reynolds & P.R.O.Bally 1958

Distribution: Somalia.

Aloe broomii Schönland 1907

Distribution: Lesotho, South Africa.

Aloe broomii var. **broomii**

Distribution: Lesotho, South Africa (Eastern Cape, Western Cape, Northern Cape, Free State).

Aloe broomii var. **tarkaensis** Reynolds 1936

Distribution: South Africa (Eastern Cape).

Aloe brunneodentata Lavranos & Collen. 2000

Distribution: Saudi Arabia

Aloe brunneostriata Lavranos & S.Carter 1992

Distribution: Somalia.

Aloe buchananii Baker 1895

Distribution: Malawi.

Aloe buchlohii Rauh 1966

Distribution: Madagascar.

Aloe buettneri A.Berger 1905
Aloe barteri Baker 1880
Aloe barteri var. *dahomensis* A.Chev. 1952 (nom. nud., Art. 36.1)
Aloe barteri var. *sudanica* A.Chev. 1952 (nom. nud., Art. 36.1)
Aloe paludicola A.Chev. 1952 (nom. nud., Art. 36.1)

Distribution: Benin, Ghana, Mali, Nigeria, Togo.

Aloe buhrii Lavranos 1971

Distribution: South Africa (Northern Cape).

Part II: Aloe

Aloe bukobana Reynolds 1955

Distribution: Tanzania (United Republic of).

Aloe bulbicaulis Christian 1936
Aloe trothae A.Berger 1905

Distribution: Angola, Democratic Republic of the Congo (the), Malawi, Tanzania (United Republic of), Zambia.

Aloe bulbillifera H.Perrier 1926

Distribution: Madagascar.

Aloe bulbillifera var. **bulbillifera**

Distribution: Madagascar.

Aloe bulbillifera var. **paulianae** Reynolds 1956

Distribution: Madagascar.

Aloe bullockii Reynolds 1961

Distribution: Tanzania (United Republic of).

Aloe burgersfortensis Reynolds 1936

Distribution: South Africa (Mpumalanga).

Aloe bussei A.Berger 1908
Aloe morogoroensis Christian 1940

Distribution: Tanzania (United Republic of).

Aloe calcairophila Reynolds 1961
Aloe calcairophylla hort. (nom. nud., Art. 61.1)
Guillauminia calcairophila (Reynolds) P.V.Heath 1994

Distribution: Madagascar.

Aloe calidophila Reynolds 1954

Distribution: Ethiopia, Kenya.

Aloe cameronii Hemsl. 1903

Distribution: Malawi, Mozambique, Zambia, Zimbabwe.

Aloe cameronii var. **bondana** Reynolds 1966

Distribution: Zimbabwe.

Aloe cameronii var. **cameronii**

Distribution: Malawi, Mozambique, Zambia, Zimbabwe.

Aloe cameronii var. **dedzana** Reynolds 1965

Distribution: Malawi, Mozambique.

Aloe camperi Schweinf. 1894
Aloe abyssinica Salm-Dyck 1817 (nom. illeg., Art. 53.1)
Aloe albopicta hort. ex A.Berger 1908
Aloe eru A.Berger 1908
Aloe eru forma *erecta* hort. ex A.Berger 1908
Aloe eru forma *glauca* hort. ex A.Berger 1908
Aloe eru forma *maculata* hort. ex A.Berger 1908
Aloe eru forma *parvipunctata* hort. ex A.Berger 1908
Aloe eru var. *cornuta* A.Berger 1908
Aloe spicata Baker 1896 (nom. illeg., Art. 53.1)

Distribution: Eritrea, Ethiopia.

Aloe canarina S.Carter 1994

Distribution: Sudan (the), Uganda.

Aloe cannellii L.C.Leach 1971

Distribution: Mozambique.

Aloe capitata Baker 1883

Distribution: Madagascar.

Aloe capitata var. **capitata**
Aloe cernua Tod. 1890

Distribution: Madagascar.

Aloe capitata var. **cipolinicola** H.Perrier 1926

Distribution: Madagascar.

Aloe capitata var. **gneissicola** H.Perrier 1926

Distribution: Madagascar.

Part II: Aloe

Aloe capitata var. **quartziticola** H.Perrier 1926

Distribution: Madagascar.

Aloe capitata var. **silvicola** H.Perrier 1926

Distribution: Madagascar.

Aloe capmanambatoensis Rauh & Gerold 2000

Distribution: Madagascar (NE)

Aloe carnea S.Carter 1996

Distribution: Zimbabwe.

Aloe castanea Schönland 1907

Distribution: South Africa (Gauteng, Mpumalanga, Northern Prov.).

Aloe castellorum J.R.I.Wood 1983

Distribution: Saudi Arabia, Yemen.

Aloe catengiana Reynolds 1961

Distribution: Angola.

Aloe cephalophora Lavranos & Collen. 2000

Distribution: Saudi Arabia

Aloe chabaudii Schönland 1905

Distribution: Democratic Republic of the Congo (the), Malawi, Mozambique, South Africa, Swaziland, Tanzania (United Republic of), Zambia, Zimbabwe.

Aloe chabaudii var. **chabaudii**

Distribution: Democratic Republic of the Congo (the), Malawi, Mozambique, South Africa, Swaziland, Tanzania (United Republic of), Zambia, Zimbabwe.

Aloe chabaudii var. **mlanjeana** Christian 1938

Distribution: Malawi.

Aloe chabaudii var. **verekeri** Christian 1938

Distribution: Mozambique, Zimbabwe.

Aloe cheranganiensis S.Carter & Brandham 1979

Distribution: Kenya, Uganda.

Aloe chlorantha Lavranos 1973

Distribution: South Africa (Northern Cape).

Aloe chortolirioides A.Berger 1908

Distribution: South Africa, Swaziland.

Aloe chortolirioides var. **chortolirioides**
Aloe boastii Letty 1934
Aloe chortolirioides var. *boastii* (Letty) Reynolds 1950

Distribution: South Africa (Mpumalanga), Swaziland.

Aloe chortolirioides var. **woolliana** (Pole-Evans) Glen & D.S.Hardy 1987
Aloe woolliana Pole-Evans 1934

Distribution: South Africa (Mpumalanga, Northern Prov.), Swaziland.

Aloe christianii Reynolds 1936

Distribution: Angola, Democratic Republic of the Congo (the), Malawi, Mozambique, Tanzania (United Republic of), Zambia, Zimbabwe.

Aloe chrysostachys Lavranos & L.E.Newton 1976
Aloe meruana Lavranos 1980

Distribution: Kenya.

Aloe ciliaris Haw. 1825

Distribution: South Africa.

Aloe ciliaris var. **ciliaris**
Aloe ciliaris forma *flanaganii* (Schönland) Resende 1943
Aloe ciliaris forma *gigas* Resende 1938
Aloe ciliaris var. *flanaganii* Schönland 1903

Distribution: South Africa (Eastern Cape).

Part II: Aloe

Aloe ciliaris var. **redacta** S.Carter 1990

Distribution: South Africa (Eastern Cape).

Aloe ciliaris var. **tidmarshii** Schönland 1903
Aloe ciliaris forma *tidmarshii* (Schönland) Resende 1943
Aloe tidmarshii (Schönland) F.S.Mull. ex R.A.Dyer 1943

Distribution: South Africa (Eastern Cape).

Aloe citrea (Guillaumin) L.E.Newton & G.D.Rowley 1997
Lomatophyllum citreum Guillaumin 1944

Distribution: Madagascar.

Aloe citrina S.Carter & Brandham 1983

Distribution: Ethiopia, Kenya, Somalia.

Aloe classenii Reynolds 1965

Distribution: Kenya.

Aloe claviflora Burch. 1822
Aloe decora Schönland 1905
Aloe schlechteri Schönland 1903

Distribution: Namibia, South Africa (Northern Cape, Western Cape, Eastern Cape, Free State).

Aloe collenetteae Lavranos 1995

Distribution: Oman (Dhofar).

Aloe collina S.Carter 1996

Distribution: Zimbabwe.

Aloe commixta A.Berger 1908
Aloe gracilis Baker 1880 (nom. illeg., Art. 53.1)
Aloe perfoliata var. α L. 1753

Distribution: South Africa (Western Cape).

Aloe comosa Marloth & A.Berger 1905

Distribution: South Africa (Western Cape).

Aloe compressa H.Perrier 1926

Distribution: Madagascar.

Aloe compressa var. **compressa**

Distribution: Madagascar.

Aloe compressà var. **paucituberculata** Lavranos 1998

Distribution: Madagascar (C).

Aloe compressa var. **rugosquamosa** H.Perrier 1926

Distribution: Madagascar.

Aloe compressa var. **schistophila** H.Perrier 1926

Distribution: Madagascar.

Aloe comptonii Reynolds 1950

Distribution: South Africa (Western Cape, Eastern Cape).

Aloe confusa Engl. 1895

Distribution: Kenya, Tanzania (United Republic of).

Aloe congdonii S.Carter 1994

Distribution: Tanzania (United Republic of).

Aloe conifera H.Perrier 1926

Distribution: Madagascar.

Aloe cooperi Baker 1874

Distribution: South Africa, Swaziland.

Aloe cooperi ssp. **cooperi**
Aloe schmidtiana Regel 1879

Distribution: South Africa (KwaZulu-Natal, Mpumalanga), Swaziland.

Part II: Aloe

Aloe cooperi ssp. **pulchra** Glen & D.S.Hardy 1987

Distribution: South Africa (KwaZulu-Natal), Swaziland.

Aloe corallina I.Verd. 1979

Distribution: Namibia.

Aloe crassipes Baker 1880

Distribution: Sudan (the), Zambia.

Aloe cremersii Lavranos 1974

Distribution: Madagascar.

Aloe cremnophila Reynolds & P.R.O.Bally 1961

Distribution: Somalia.

Aloe cryptoflora Reynolds 1965

Distribution: Madagascar.

Aloe cryptopoda Baker 1884
Aloe pienaarii Pole-Evans 1915
Aloe wickensii Pole-Evans 1915
Aloe wickensii var. *lutea* Reynolds 1935
Aloe wickensii var. *wickensii*

Distribution: Botswana, Malawi, Mozambique, South Africa (Northern Prov., North-West Prov., Mpumalanga), Swaziland, Zambia, Zimbabwe.

Aloe cyrtophylla Lavranos 1998

Distribution: Madagascar (S).

Aloe dabenorisana Van Jaarsv. 1982

Distribution: South Africa (Northern Cape).

Aloe dawei A.Berger 1906
Aloe beniensis De Wild. 1921
Aloe pole-evansii Christian 1940

Distribution: Democratic Republic of the Congo (the), Kenya, Rwanda, Uganda.

Aloe debrana Christian 1947
Aloe berhana Reynolds 1957

Distribution: Ethiopia.

Aloe decorsei H.Perrier 1926

Distribution: Madagascar.

Aloe decurva Reynolds 1957

Distribution: Mozambique.

Aloe delphinensis Rauh 1990

Distribution: Madagascar.

Aloe deltoideodonta Baker 1883

Distribution: Madagascar.

Aloe deltoideodonta var. **brevifolia** H.Perrier 1926

Distribution: Madagascar.

Aloe deltoideodonta var. **candicans** H.Perrier 1926

Distribution: Madagascar.

Aloe deltoideodonta var. **deltoideodonta**
Aloe deltoideodonta var. *typica* H.Perrier 1926 (nom. nud., Art. 24.3)

Distribution: Madagascar.

Aloe descoingsii Reynolds 1958
Guillauminia descoingsii (Reynolds) P.V.Heath 1994

Distribution: Madagascar.

Aloe descoingsii ssp. **augustina** Lavranos 1995

Distribution: Madagascar.

Aloe descoingsii ssp. **descoingsii**

Distribution: Madagascar.

Part II: Aloe

Aloe deserti A.Berger 1905

Distribution: Kenya, Tanzania (United Republic of).

Aloe dewetii Reynolds 1937

Distribution: South Africa (KwaZulu-Natal, Mpumalanga), Swaziland.

Aloe dewinteri Giess 1973

Distribution: Namibia.

Aloe dhufarensis Lavranos 1967

Distribution: Oman.

Aloe dichotoma Masson 1776
Aloe dichotoma var. *montana* (Schinz) A.Berger 1908
Aloe montana Schinz 1896
Aloe ramosa Haw. 1804
Rhipidodendrum dichotomum (Masson) Willd. 1811

Distribution: Namibia, South Africa (Northern Cape).

Aloe dinteri A.Berger 1914

Distribution: Namibia.

Aloe diolii L.E.Newton 1995

Distribution: Sudan (the).

Aloe distans Haw. 1812
Aloe brevifolia Haw. 1804 (nom. illeg., Art. 53.1)
Aloe mitriformis var. *angustior* Lam. 1784
Aloe mitriformis var. *brevifolia* Aiton 1810
Aloe perfoliata var. *brevifolia* Aiton 1789

Distribution: South Africa (Western Cape).

Aloe divaricata A.Berger 1905

Distribution: Madagascar.

Aloe divaricata var. **divaricata**
Aloe sahundra Bojer 1837 (nom. nud., Art. 32.1c)
Aloe vahontsohy Decorse ex Poiss. 1912
Aloe vahontsohy H.Perrier 1938 (nom. nud.)

Aloe vaotsohy Decorse & Poiss. 1912

Distribution: Madagascar.

Aloe divaricata var. **rosea** (Decary) Reynolds 1958
Aloe vaotsohy var. *rosea* Decary 1921

Distribution: Madagascar.

Aloe doei Lavranos 1965

Distribution: Yemen.

Aloe doei var. **doei**

Distribution: Yemen.

Aloe doei var. **lavranosii** Marn.-Lap. 1970

Distribution: Yemen.

Aloe dominella Reynolds 1938

Distribution: South Africa (KwaZulu-Natal).

Aloe dorotheae A.Berger 1908
Aloe harmsii A.Berger 1908

Distribution: Tanzania (United Republic of).

Aloe duckeri Christian 1940

Distribution: Malawi, Tanzania (United Republic of), Zambia.

Aloe dyeri Schönland 1905

Distribution: South Africa (Mpumalanga).

Aloe ecklonis Salm-Dyck 1849

Distribution: Lesotho, South Africa (Eastern Cape, KwaZulu-Natal, Free State, Mpumalanga), Swaziland.

Aloe edentata Lavranos & Collen.

Distribution: Saudi Arabia

Part II: Aloe

Aloe elata S.Carter & L.E.Newton 1994

Distribution: Kenya, Tanzania (United Republic of).

Aloe elegans Tod. 1882
Aloe abyssinica A.Berger 1908 (nom. illeg., Art. 53.1)
Aloe abyssinica var. *peacockii* Baker 1880
Aloe aethiopica (Schweinf.) A.Berger 1905
Aloe peacockii A.Berger 1905
Aloe percrassa var. *saganeitiana* A.Berger 1908
Aloe schweinfurthii Hook.f. 1899 (nom. illeg., Art. 53.1)
Aloe vera var. *aethiopica* Schweinf. 1894

Distribution: Eritrea, Ethiopia.

Aloe elgonica Bullock 1932

Distribution: Kenya.

Aloe ellenbeckii A.Berger 1905
Aloe dumetorum B.Mathew & Brandham 1977

Distribution: Ethiopia, Kenya, Somalia.

Aloe eminens Reynolds & P.R.O.Bally 1958

Distribution: Somalia.

Aloe enotata L.C.Leach 1972

Distribution: Zambia.

Aloe eremophila Lavranos 1965

Distribution: Yemen (Hadhramaut).

Aloe erensii Christian 1940

Distribution: Kenya, Sudan (the).

Aloe ericetorum Bosser 1968

Distribution: Madagascar (C).

Aloe erythrophylla Bosser 1968

Distribution: Madagascar (C).

Aloe esculenta L.C.Leach 1971

Distribution: Angola, Botswana, Namibia, Zambia.

Aloe eumassawana S.Carter & al. 1996

Distribution: Eritrea.

Aloe excelsa A.Berger 1906

Distribution: Malawi, Mozambique, South Africa, Zambia, Zimbabwe.

Aloe excelsa var. **breviflora** L.C.Leach 1977

Distribution: Malawi, Mozambique.

Aloe excelsa var. **excelsa**

Distribution: Malawi, Mozambique, South Africa (Northern Prov.), Zambia, Zimbabwe.

Aloe falcata Baker 1880

Distribution: South Africa (Northern Cape, Western Cape).

Aloe ferox Mill. 1768
Aloe candelabrum A.Berger 1906
Aloe ferox var. *erythrocarpa* A.Berger 1908
Aloe ferox var. *galpinii* (Baker) Reynolds 1937
Aloe ferox var. *hanburyi* Baker 1896
Aloe ferox var. *incurva* Baker 1880
Aloe ferox var. *subferox* (Spreng.) Baker 1880
Aloe galpinii Baker 1901
Aloe muricata Haw. 1804
Aloe perfoliata Thunb. 1785
Aloe perfoliata var. *ferox* Aiton 1789
Aloe perfoliata var. γ L. 1753
Aloe perfoliata var. ε L. 1753
Aloe perfoliata var. ζ Willd. 1799
Aloe pseudoferox Salm-Dyck 1817
Aloe socotorina Masson 1773
Aloe subferox Spreng. 1826
Aloe supralaevis Haw. 1804
Pachidendron ferox (Mill.) Haw. 1821
Pachidendron pseudoferox (Salm-Dyck) Haw. 1821
Pachidendron supralaeve (Haw.) Haw. 1821

Distribution: Lesotho, South Africa (Western Cape, Eastern Cape, Free State, KwaZulu-Natal).

Aloe fibrosa Lavranos & L.E.Newton 1976

Distribution: Kenya, Tanzania (United Republic of).

Part II: Aloe

Aloe fievetii Reynolds 1965

Distribution: Madagascar.

Aloe fimbrialis S.Carter 1996

Distribution: Tanzania (United Republic of), Zambia.

Aloe fleurentiniorum Lavranos & L.E.Newton 1977

Distribution: Yemen.

Aloe fleuretteana Rauh & Gerold 2000

Distribution: Madagascar.

Aloe flexilifolia Christian 1942

Distribution: Tanzania (United Republic of).

Aloe forbesii Balf.f. 1903

Distribution: Yemen (Socotra).

Aloe fosteri Pillans 1933

Distribution: South Africa (Mpumalanga).

Aloe fouriei D.S.Hardy & Glen 1987

Distribution: South Africa (Mpumalanga).

Aloe fragilis Lavranos & Röösli 1994

Distribution: Madagascar (N).

Aloe framesii L.Bolus 1933
Aloe amoena Pillans 1933
Aloe microstigma ssp. *framesii* (L.Bolus) Glen & D.S.Hardy

Distribution: South Africa (Northern Cape, Western Cape).

Aloe francombei L.E.Newton 1994

Distribution: Kenya.

Aloe friisii Sebsebe & M.G.Gilbert 2000

Distribution: Ethiopia

Aloe fulleri Lavranos 1967

Distribution: Yemen.

Aloe gariepensis Pillans 1933
Aloe gariusiana Dinter 1928 (nom. nud., Art. 32.1c)

Distribution: Namibia, South Africa (Northern Cape).

Aloe gerstneri Reynolds 1937

Distribution: South Africa (KwaZulu-Natal).

Aloe gilbertii T.Reynolds ex Sebsebe & Brandham 1992

Distribution: Ethiopia.

Aloe gilbertii ssp. **gilbertii**

Distribution: Ethiopia.

Aloe gilbertii ssp. **megalacanthoides** M.G.Gilbert & Sebsebe 1997

Distribution: Ethiopia.

Aloe gillettii S.Carter 1994

Distribution: Somalia.

Aloe glabrescens (Reynolds & P.R.O.Bally) S.Carter & Brandham 1983
Aloe rigens var. *glabrescens* Reynolds & P.R.O.Bally 1958

Distribution: Somalia.

Aloe glauca Mill. 1978

Distribution: South Africa.

Aloe glauca var. **glauca**
Aloe glauca var. *elatior* Salm-Dyck 1817
Aloe glauca var. *humilior* Salm-Dyck 1817
Aloe glauca var. *major* Haw. 1812
Aloe glauca var. *minor* Haw. 1812
Aloe perfoliata var. *glauca* Aiton 1789

Part II: Aloe

Aloe perfoliata var. κ L. 1753
Aloe rhodacantha DC. 1799

Distribution: South Africa (Northern Cape, Western Cape).

Aloe glauca var. **spinosior** Haw. 1821
Aloe glauca var. *muricata* (Schult.) Baker 1880
Aloe muricata Schult. 1809

Distribution: South Africa (Northern Cape, Western Cape).

Aloe globuligemma Pole-Evans 1915

Distribution: Botswana, South Africa (Northern Prov., Mpumalanga), Zimbabwe.

Aloe gossweileri Reynolds 1962

Distribution: Angola.

Aloe gracilicaulis Reynolds & P.R.O.Bally 1958

Distribution: Somalia.

Aloe gracilis Haw. 1825

Distribution: South Africa.

Aloe gracilis var. **decumbens** Reynolds 1950

Distribution: South Africa (Western Cape).

Aloe gracilis var. **gracilis**
Aloe laxiflora N.E.Br. 1906

Distribution: South Africa (Eastern Cape).

Aloe grandidentata Salm-Dyck 1822

Distribution: Botswana, South Africa (Northern Cape, North-West Prov., Free State).

Aloe grata Reynolds 1960

Distribution: Angola.

Aloe greatheadii Schönland 1904

Distribution: Botswana, Democratic Republic of the Congo (the), Malawi, Mozambique, South Africa, Swaziland, Zimbabwe.

Aloe greatheadii var. **davyana** (Schönland) Glen & D.S.Hardy 1987
Aloe comosibracteata Reynolds 1936
Aloe davyana Schönland 1904
Aloe davyana var. *subolifera* Groenew. 1939
Aloe graciliflora Groenew. 1936
Aloe labiaflava Groenew. 1936
Aloe longibracteata Pole-Evans 1915
Aloe mutans Reynolds 1936
Aloe verdoorniae Reynolds 1936

Distribution: South Africa (Free State, Gauteng, KwaZulu-Natal, Northern Prov., North-West Prov.), Swaziland.

Aloe greatheadii var. **greatheadii**
Aloe pallidiflora A.Berger 1905
Aloe termetophila De Wild. 1921

Distribution: Botswana, Democratic Republic of the Congo (the), Malawi, Mozambique, South Africa (Northern Prov.), Zimbabwe.

Aloe greenii Baker 1880

Distribution: South Africa (KwaZulu-Natal).

Aloe grisea S.Carter & Brandham 1983

Distribution: Somalia.

Aloe guerrae Reynolds 1960

Distribution: Angola.

Aloe guillaumetii Cremers 1976

Distribution: Madagascar.

Aloe haemanthifolia A.Berger & Marloth 1905

Distribution: South Africa (Western Cape).

Aloe hardyi Glen 1987

Distribution: South Africa (Mpumalanga).

Aloe harlana Reynolds 1957

Distribution: Ethiopia.

Part II: Aloe

Aloe haworthioides Baker 1886
Aloinella haworthioides (Baker) Lemée 1939 (nom. nud., Art. 43.1)
Lemeea haworthioides (Baker) P.V.Heath 1993

Distribution: Madagascar.

Aloe haworthioides var. **aurantiaca** H.Perrier 1926

Distribution: Madagascar.

Aloe haworthioides var. **haworthioides**

Distribution: Madagascar.

Aloe hazeliana Reynolds 1959

Distribution: Mozambique, Zimbabwe.

Aloe helenae Danguy 1929

Distribution: Madagascar (S).

Aloe heliderana Lavranos 1973

Distribution: Somalia.

Aloe hemmingii Reynolds & P.R.O.Bally 1964

Distribution: Somalia.

Aloe hendrickxii Reynolds 1955

Distribution: Democratic Republic of the Congo (the).

Aloe hereroensis Engl. 1888
Aloe hereroensis var. *hereroensis*

Distribution: Angola, Namibia, South Africa (Northern Cape).

Aloe heybensis Lavranos 1999

Distribution: Somalia (S).

Aloe hijazensis Lavranos & Collen. 2000

Distribution: Saudi Arabia

Aloe hildebrandtii Baker 1888
Aloe gloveri Reynolds & P.R.O.Bally 1958

Distribution: Somalia.

Aloe hlangapies Groenew. 1936
Aloe hlangapensis Groenew. 1937 (nom. nud., Art. 61.1)
Aloe hlangapitis Groenew. 1936 (nom. nud., Art. 61.1)

Distribution: South Africa (KwaZulu-Natal, Mpumalanga).

Aloe howmanii Reynolds 1961

Distribution: Zimbabwe.

Aloe humbertii H.Perrier 1931

Distribution: Madagascar (S).

Aloe humilis (L.) Mill. 1771
Aloe acuminata (Haw.) Haw. 1812
Aloe acuminata var. *major* Salm-Dyck 1817
Aloe echinata Willd. 1809
Aloe humilis Ker Gawl. 1804 (nom. illeg., Art. 53.1)
Aloe humilis subvar. *minor* Salm-Dyck 1837
Aloe humilis subvar. *semiguttata* Haw. 1821
Aloe humilis var. *acuminata* (Haw.) Baker 1880
Aloe humilis var. *candollei* Baker 1880
Aloe humilis var. *echinata* (Willd.) Baker 1896
Aloe humilis var. *humilis*
Aloe humilis var. *incurva* Haw. 1804
Aloe humilis var. *suberecta* (Haw.) Baker 1896
Aloe humilis var. *subtuberculata* (Haw.) Baker 1896
Aloe incurva (Haw.) Haw. 1812
Aloe perfoliata var. *humilis* L. 1753
Aloe suberecta Haw. 1804
Aloe suberecta var. *acuminata* Haw. 1804
Aloe subtuberculata Haw. 1825
Aloe tuberculata Haw. 1804
Catevala humilis (L.) Medik. 1786

Distribution: South Africa (Western Cape, Eastern Cape).

Aloe ibitiensis H.Perrier 1926
Aloe ibityensis hort. (nom. nud., Art. 61.1)

Distribution: Madagascar (C).

Aloe imalotensis Reynolds 1957
Aloe contigua (H.Perrier) Reynolds 1958
Aloe deltoideodonta forma *latifolia* H.Perrier 1938 (nom. nud., Art. 36.1)
Aloe deltoideodonta forma *longifolia* H.Perrier 1938 (nom. nud., Art. 36.1)
Aloe deltoideodonta subforma *variegata* Boiteau ex H.Jacobsen 1954 (nom. nud., Art. 36.1)

Part II: Aloe

Aloe deltoideodonta var. *contigua* H.Perrier 1926

Distribution: Madagascar.

Aloe ×imerinensis Bosser 1968

Distribution: Madagascar.

Aloe immaculata Pillans 1934

Distribution: South Africa (Northern Prov.).

Aloe inamara L.C.Leach 1971

Distribution: Angola.

Aloe inconspicua Plowes 1986

Distribution: South Africa (KwaZulu-Natal).

Aloe inermis Forssk. 1775

Distribution: Yemen.

Aloe integra Reynolds 1936

Distribution: South Africa (Mpumalanga), Swaziland.

Aloe inyangensis Christian 1936

Distribution: Zimbabwe.

Aloe inyangensis var. **inyangensis**

Distribution: Zimbabwe.

Aloe inyangensis var. **kimberleyana** S.Carter 1996

Distribution: Zimbabwe.

Aloe isaloensis H.Perrier 1927

Distribution: Madagascar.

Aloe itremensis Reynolds 1955

Distribution: Madagascar.

Aloe jacksonii Reynolds 1955

Distribution: Ethiopia (Ogaden).

Aloe jucunda Reynolds 1953

Distribution: Somalia.

Aloe juvenna Brandham & S.Carter 1979

Distribution: Kenya.

Aloe ×keayi Reynolds (pro sp.) 1963

Distribution: Ghana.

Aloe kedongensis Reynolds 1953
Aloe nyeriensis ssp. *kedongensis* (Reynolds) S.Carter 1980

Distribution: Kenya.

Aloe kefaensis M.G.Gilbert & Sebsebe 1997

Distribution: Ethiopia.

Aloe keithii Reynolds 1937

Distribution: Swaziland.

Aloe ketabrowniorum L.E.Newton 1994

Distribution: Kenya.

Aloe khamiesensis Pillans 1934

Distribution: South Africa (Northern Cape).

Aloe kilifiensis Christian 1942

Distribution: Kenya.

Aloe kniphofioides Baker 1890
Aloe marshallii J.M.Wood & M.S.Evans 1897

Distribution: South Africa (KwaZulu-Natal, Mpumalanga).

Part II: Aloe

Aloe krapohliana Marloth 1908

Distribution: South Africa.

Aloe krapohliana var. **dumoulinii** Lavranos 1973

Distribution: South Africa (Northern Cape).

Aloe krapohliana var. **krapohliana**

Distribution: South Africa (Northern Cape, Western Cape).

Aloe kraussii Baker 1880

Distribution: South Africa (KwaZulu-Natal).

Aloe kulalensis L.E.Newton & Beentje 1990

Distribution: Kenya (N).

Aloe labworana (Reynolds) S.Carter 1994
Aloe schweinfurthii var. *labworana* Reynolds 1956

Distribution: Sudan (the), Uganda.

Aloe laeta A.Berger 1908

Distribution: Madagascar.

Aloe laeta var. **laeta**

Distribution: Madagascar.

Aloe laeta var. **maniaensis** H.Perrier 1926

Distribution: Madagascar.

Aloe lateritia Engl. 1895

Distribution: Ethiopia, Kenya, Tanzania (United Republic of).

Aloe lateritia var. **graminicola** (Reynolds) S.Carter 1994
Aloe graminicola Reynolds 1953
Aloe solaiana Christian 1940

Distribution: Ethiopia, Kenya.

Aloe lateritia var. **lateritia**
Aloe amanensis A.Berger 1905
Aloe boehmii Engl. 1895
Aloe campylosiphon A.Berger 1904

Distribution: Kenya, Tanzania (United Republic of).

Aloe lavranosii Reynolds 1964

Distribution: Yemen.

Aloe leachii Reynolds 1965

Distribution: Tanzania (United Republic of).

Aloe leandrii Bosser 1968

Distribution: Madagascar.

Aloe leedalii S.Carter 1994

Distribution: Tanzania (United Republic of).

Aloe lensayuensis Lavranos & L.E.Newton 1976

Distribution: Kenya.

Aloe lepida L.C.Leach 1974

Distribution: Angola.

Aloe leptosiphon A.Berger 1905
Aloe greenwayi Reynolds 1964

Distribution: Tanzania (United Republic of).

Aloe lettyae Reynolds 1937

Distribution: South Africa (Northern Prov.).

Aloe lindenii Lavranos 1997

Distribution: Somalia.

Aloe linearifolia A.Berger 1922

Distribution: South Africa (KwaZulu-Natal).

Part II: Aloe

Aloe lineata (Aiton) Haw. 1804
Aloe perfoliata var. *lineata* Aiton 1789

Distribution: South Africa.

Aloe lineata var. **lineata**
Aloe lineata var. *glaucescens* Haw. 1821
Aloe lineata var. *viridis* Haw. 1821

Distribution: South Africa (Western Cape, Eastern Cape).

Aloe lineata var. **muirii** (Marloth) Reynolds 1950
Aloe muirii Marloth 1929

Distribution: South Africa (Western Cape).

Aloe littoralis Baker 1878
Aloe rubrolutea Schinz 1896
Aloe schinzii Baker 1898

Distribution: Angola, Botswana, Mozambique, Namibia, South Africa, Zambia, Zimbabwe.

Aloe lolwensis L.E.Newton

Distribution: Kenya

Aloe lomatophylloides Balf.f. 1877
Lomatophyllum lomatophylloides (Balf.f.) Marais 1975

Distribution: Mauritius (Rodrigues).

Aloe longistyla Baker 1880

Distribution: South Africa (Western Cape, Eastern Cape).

Aloe luapulana L.C.Leach 1972

Distribution: Zambia.

Aloe lucile-allorgeae Rauh 1998

Distribution: Madagascar.

Aloe luntii Baker 1894

Distribution: Oman, Somalia, Yemen.

Aloe lutescens Groenew. 1938

Distribution: South Africa (Northern Prov.).

Aloe macleayi Reynolds 1955

Distribution: Sudan (the).

Aloe macra Haw. 1819
Lomatophyllum macrum (Haw.) Salm-Dyck ex Roem. & Schult. 1829
Phylloma macrum (Haw.) Sweet 1827

Distribution: Réunion (French).

Aloe macrocarpa Tod. 1875
Aloe barteri Schnell 1953 (nom. illeg., Art. 53.1)
Aloe commutata Engl. 1892
Aloe edulis A.Chev. 1920
Aloe macrocarpa var. *major* A.Berger 1908

Distribution: Benin, Cameroon, Djibouti, Eritrea, Ethiopia, Ghana, Mali, Nigeria, Sudan (the).

Aloe macroclada Baker 1883

Distribution: Madagascar.

Aloe macrosiphon Baker 1898
Aloe compacta Reynolds 1961
Aloe mwanzana Christian 1940

Distribution: Kenya, Rwanda, Tanzania (United Republic of), Uganda.

Aloe maculata All. 1773
Aloe disticha Mill. 1768 (nom. illeg., Art. 53.1)
Aloe latifolia Haw. 1812
Aloe leptophylla N.E.Br. ex Baker 1880
Aloe leptophylla var. *stenophylla* Baker 1896
Aloe maculata Medik. 1786 (nom. illeg., Art. 53.1)
Aloe maculosa Lam. 1783
Aloe perfoliata var. δ L. 1753
Aloe perfoliata var. θ L. 1753
Aloe perfoliata var. λ L. 1753
Aloe perfoliata var. *saponaria* Aiton 1789
Aloe saponaria (Aiton) Haw. 1804
Aloe saponaria var. *brachyphylla* Baker 1880
Aloe saponaria var. *ficksburgensis* Reynolds 1937
Aloe saponaria var. *latifolia* Haw. 1804
Aloe saponaria var. *saponaria*
Aloe umbellata DC. 1799

Distribution: Lesotho, South Africa (Eastern Cape, Free State, KwaZulu-Natal, Mpumalanga, Western Cape), Swaziland.

Part II: Aloe

Aloe madecassa H.Perrier 1926

Distribution: Madagascar.

Aloe madecassa var. **lutea** Guillaumin 1955

Distribution: Madagascar.

Aloe madecassa var. **madecassa**

Distribution: Madagascar.

Aloe marlothii A.Berger 1905
Aloe ferox A.Berger 1908 (nom. illeg., Art. 53.1)
Aloe ferox var. *xanthostachys* A.Berger 1908 (incorrect name, Art. 11.4)
Aloe marlothii J.M.Wood 1912 (nom. illeg., Art. 53.1)

Distribution: Botswana, Mozambique, South Africa, Swaziland.

Aloe marlothii ssp. **orientalis** Glen & D.S.Hardy 1987

Distribution: Mozambique, South Africa (KwaZulu-Natal).

Aloe marlothii var. **bicolor** Reynolds 1936

Distribution: South Africa.

Aloe marlothii var. **marlothii**
Aloe spectabilis Reynolds 1927
Aloe supralaevis var. *hanburyi* Baker 1896

Distribution: Botswana, Mozambique, South Africa (KwaZulu-Natal, Gauteng, Mpumalanga, Northern Prov., Northwest Prov.), Swaziland.

Aloe massawana Reynolds 1959
Aloe kirkii Baker 1894

Distribution: Tanzania (United Republic of).

Aloe mawii Christian 1940

Distribution: Malawi, Mozambique, Tanzania (United Republic of).

Aloe mayottensis A.Berger 1908

Distribution: Comoros (the).

Aloe mccoyi Lavranos & Mies 2001

Distribution: Yemen

Aloe mcloughlinii Christian 1951
Aloe maclaughlinii hort. (nom. nud., Art. 61.1)

Distribution: Djibouti, Ethiopia.

Aloe medishiana Reynolds & P.R.O.Bally 1958

Distribution: Somalia.

Aloe megalacantha Baker 1898

Distribution: Ethiopia, Somalia.

Aloe megalacantha ssp. **alticola** M.G.Gilbert & Sebsebe 1997

Distribution: Ethiopia.

Aloe megalacantha ssp. **megalacantha**
Aloe magnidentata I.Verd. & Christian 1947

Distribution: Ethiopia, Somalia.

Aloe megalocarpa Lavranos 1998

Distribution: Madagascar (N).

Aloe melanacantha A.Berger 1905

Distribution: Namibia, South Africa.

Aloe melanacantha var. **erinacea** (D.S.Hardy) G.D.Rowley 1980
Aloe erinacea D.S.Hardy 1971

Distribution: Namibia.

Aloe melanacantha var. **melanacantha**

Distribution: Namibia, South Africa (Northern Cape).

Aloe menachensis (Schweinf.) Blatt. 1936
Aloe percrassa var. *menachensis* Schweinf. 1894
Aloe trichosantha var. *menachensis* (Schweinf.) A.Berger 1908

Distribution: Yemen.

Part II: Aloe

Aloe mendesii Reynolds 1964

Distribution: Angola, Namibia.

Aloe menyharthii Baker 1898

Distribution: Malawi, Mozambique.

Aloe menyharthii ssp. **ensifolia** S.Carter 1996

Distribution: Mozambique.

Aloe menyharthii ssp. **menyharthii**

Distribution: Malawi, Mozambique.

Aloe metallica Engl. & Gilg 1908

Distribution: Angola.

Aloe meyeri Van Jaarsv. 1981
Aloe richtersveldensis Venter & Beukes 1982

Distribution: Namibia, South Africa (Northern Cape).

Aloe micracantha Haw. 1819
Aloe micracantha Link & Otto 1825 (nom. illeg., Art. 53.1)

Distribution: South Africa (Eastern Cape).

Aloe microdonta Chiov. 1928

Distribution: Kenya, Somalia.

Aloe microstigma Salm-Dyck 1854
Aloe brunnthaleri A.Berger ex Cammerl. 1933
Aloe juttae Dinter 1923

Distribution: South Africa (Western Cape, Eastern Cape).

Aloe millotii Reynolds 1956

Distribution: Madagascar (Toliara).

Aloe milne-redheadii Christian 1940

Distribution: Angola, Zambia.

Aloe minima Baker 1895
Leptaloe minima (Baker) Stapf 1933

Distribution: South Africa, Swaziland.

Aloe minima var. **blyderivierensis** (Groenew.) Reynolds 1947
Leptaloe blyderivierensis Groenew. 1938

Distribution: South Africa (Mpumalanga).

Aloe minima var. **minima**
Aloe parviflora Baker 1901

Distribution: South Africa (KwaZulu-Natal), Swaziland.

Aloe mitriformis Mill. 1768
Aloe mitriformis var. *elatior* Haw. 1804
Aloe mitriformis var. *humilior* Haw.
Aloe parvispina Schönland 1905
Aloe perfoliata var. v L. 1753
Aloe perfoliata var. ξ Willd. 1799
Aloe perfoliata var. *mitriformis* Aiton 1789
Aloe xanthacantha Salm-Dyck 1854

Distribution: South Africa (Western Cape).

Aloe modesta Reynolds 1956

Distribution: South Africa (Mpumalanga).

Aloe molederana Lavranos & Glen 1989

Distribution: Somalia.

Aloe monotropa I.Verd. 1961

Distribution: South Africa (Northern Prov.).

Aloe monticola Reynolds 1957

Distribution: Ethiopia.

Aloe morijensis S.Carter & Brandham 1979

Distribution: Kenya, Tanzania (United Republic of).

Aloe mubendiensis Christian 1942

Distribution: Uganda.

Part II: Aloe

Aloe mudenensis Reynolds 1937

Distribution: South Africa (KwaZulu-Natal).

Aloe multicolor L.E.Newton 1994

Distribution: Kenya.

Aloe munchii Christian 1951

Distribution: Mozambique, Zimbabwe.

Aloe murina L.E.Newton 1992

Distribution: Kenya.

Aloe musapana Reynolds 1964

Distribution: Zimbabwe.

Aloe mutabilis Pillans 1933

Distribution: South Africa (Gauteng, Mpumalanga, Northern Prov., North-West Prov.).

Aloe myriacantha (Haw.) Schult. & Schult.f. 1829
Aloe caricina A.Berger 1905
Aloe graminifolia A.Berger 1905
Aloe johnstonii Baker 1887
Bowiea myriacantha Haw. 1827
Leptaloe myriacantha (Haw.) Stapf 1933

Distribution: Burundi, Democratic Republic of the Congo (the), Kenya, Malawi, Rwanda, South Africa (Eastern Cape, KwaZulu-Natal), Tanzania (United Republic of), Uganda, Zimbabwe.

Aloe mzimbana Christian 1941

Distribution: Democratic Republic of the Congo (the), Malawi, Tanzania (United Republic of), Zambia.

Aloe namibensis Giess 1970

Distribution: Namibia.

Aloe namorokaensis (Rauh) L.E.Newton & G.D.Rowley 1998
Lomatophyllum namorokaense Rauh 1998

Distribution: Madagascar (W).

Aloe ngongensis Christian 1942

Distribution: Kenya, Tanzania (United Republic of).

Aloe niebuhriana Lavranos 1965

Distribution: Yemen.

Aloe nubigena Groenew. 1936

Distribution: South Africa (Mpumalanga).

Aloe nuttii Baker 1897
Aloe brunneo-punctata Engl. & Gilg 1903
Aloe corbisieri De Wild. 1921
Aloe mketiensis Christian 1940

Distribution: Angola, Democratic Republic of the Congo (the), Malawi, Tanzania (United Republic of), Zambia.

Aloe nyeriensis Christian ex I.Verd. 1952
Aloe ngobitensis Reynolds 1953

Distribution: Kenya.

Aloe occidentalis (H.Perrier) L.E.Newton & G.D.Rowley 1997
Lomatophyllum occidentale H.Perrier 1926

Distribution: Madagascar (W).

Aloe officinalis Forssk. 1775
Aloe maculata Forssk. 1775 (nom. illeg., Art. 53.1)
Aloe officinalis var. *angustifolia* (Schweinf.) Lavranos 1965
Aloe vera var. *angustifolia* Schweinf. 1894
Aloe vera var. *officinalis* (Forssk.) Baker 1880

Distribution: Saudi Arabia, Yemen.

Aloe oligophylla Baker 1883
Lomatophyllum oligophyllum (Baker) H.Perrier 1926

Distribution: Madagascar.

Aloe orientalis (H.Perrier) L.E.Newton & G.D.Rowley 1997
Lomatophyllum orientale H.Perrier 1926

Distribution: Madagascar.

Part II: Aloe

Aloe ortholopha Christian & Milne-Redh. 1933

Distribution: Zimbabwe.

Aloe otallensis Baker 1898
Aloe boranensis Cufod. 1939

Distribution: Ethiopia.

Aloe pachygaster Dinter 1924

Distribution: Namibia.

Aloe paedogona A.Berger 1906

Distribution: Angola, Namibia.

Aloe palmiformis Baker 1878

Distribution: Angola.

Aloe parallelifolia H.Perrier 1926

Distribution: Madagascar.

Aloe parvibracteata Schönland 1907
Aloe decurvidens Groenew. 1937
Aloe komatiensis Reynolds 1936
Aloe lusitanica Groenew. 1937
Aloe parvibracteata var. *zuluensis* (Reynolds) Reynolds 1950
Aloe pongolensis Reynolds 1936
Aloe pongolensis var. *zuluensis* Reynolds 1937

Distribution: Mozambique, South Africa (KwaZulu-Natal, Mpumalanga), Swaziland.

Aloe parvicapsula Lavranos & Collen. 2000

Distribution: Saudi Arabia

Aloe parvicoma Lavranos & Collen. 2000

Distribution: Saudi Arabia

Aloe parvidens M.G.Gilbert & Sebsebe 1992

Distribution: Ethiopia, Kenya, Somalia, Tanzania (United Republic of),.

Aloe parvula A.Berger 1908
Aloe sempervivoides H.Perrier 1926
Lemeea parvula (A.Berger) P.V.Heath 1994

Distribution: Madagascar.

Aloe patersonii B.Mathew 1978

Distribution: Democratic Republic of the Congo (the).

Aloe pearsonii Schönland 1911

Distribution: Namibia, South Africa (Northern Cape).

Aloe peckii P.R.O.Bally & I.Verd. 1956

Distribution: Somalia.

Aloe peglerae Schönland 1904

Distribution: South Africa (Gauteng).

Aloe pembana L.E.Newton 1998
Lomatophyllum pembanum (L.E.Newton) Rauh 1998

Distribution: Tanzania (United Republic of) (Pemba).

Aloe pendens Forssk. 1775
Aloe arabica Lam. 1783
Aloe dependens Steud. 1840
Aloe variegata Forssk. 1775 (nom. illeg., Art. 53.1)

Distribution: Yemen.

Aloe penduliflora Baker 1888

Distribution: Kenya.

Aloe percrassa Tod. 1875
Aloe abyssinica var. *percrassa* (Tod.) Baker 1880
Aloe oligospila Baker 1902
Aloe schimperi G.Karst. & Schenk 1905 (nom. illeg., Art. 53.1)
Aloe schimperi Schweinf. 1894 (nom. illeg., Art. 53.1)

Distribution: Eritrea, Ethiopia.

Part II: Aloe

Aloe perrieri Reynolds 1956
Aloe parvula H.Perrier 1926 (nom. illeg., Art. 53.1)

Distribution: Madagascar.

Aloe perryi Baker 1881

Distribution: Yemen (Socotra).

Aloe petricola Pole-Evans 1917

Distribution: South Africa (Mpumalanga).

Aloe petrophila Pillans 1933

Distribution: South Africa (Northern Prov.).

Aloe peyrierasii Cremers 1976
Lomatophyllum peyrierasii (Cremers) Rauh 1998

Distribution: Madagascar.

Aloe pictifolia D.S.Hardy 1976

Distribution: South Africa (Eastern Cape).

Aloe pillansii L.Guthrie 1928

Distribution: Namibia, South Africa (Northern Cape).

Aloe pirottae A.Berger 1905

Distribution: Ethiopia, Kenya.

Aloe plicatilis (L.) Mill. 1768
Aloe disticha var. *plicatilis* L. 1753
Aloe flabelliformis Salisb. 1796
Aloe lingua Thunb. 1785
Aloe linguaeformis L.f. 1782
Aloe plicatilis var. *major* Salm-Dyck 1817
Aloe tripetala Medik. 1783
Kumara disticha Medik. 1786
Rhipidodendrum distichum (Medik.) Willd. 1811
Rhipidodendrum plicatile (L.) Haw. 1821

Distribution: South Africa (Western Cape).

Aloe plowesii Reynolds 1964

Distribution: Mozambique.

Aloe pluridens Haw. 1824
Aloe atherstonei Baker 1880
Aloe pluridens var. *beckeri* Schönland 1903

Distribution: South Africa (Eastern Cape, KwaZulu-Natal).

Aloe polyphylla Schönland ex Pillans 1934

Distribution: Lesotho.

Aloe porphyrostachys Lavranos & Collen. 2000

Distribution: Saudi Arabia.

Aloe powysiorum L.E.Newton & Beentje 1990

Distribution: Kenya.

Aloe pratensis Baker 1880

Distribution: Lesotho, South Africa (Eastern Cape, KwaZulu-Natal).

Aloe pretoriensis Pole-Evans 1914

Distribution: South Africa (Gauteng, Mpumalanga, Northern Prov.), Swaziland, Zimbabwe.

Aloe prinslooi I.Verd. & D.S.Hardy 1965

Distribution: South Africa (KwaZulu-Natal).

Aloe procera L.C.Leach 1974

Distribution: Angola.

Aloe propagulifera (Rauh & Razaf.) L.E.Newton & G.D.Rowley 1998
Lomatophyllum propaguliferum Rauh & Razaf. 1998

Distribution: Madagascar (C-E).

Aloe prostrata (H.Perrier) L.E.Newton & G.D.Rowley 1997
Lomatophyllum prostratum H.Perrier 1926

Distribution: Madagascar.

Part II: Aloe

Aloe prostrata ssp. **pallida** Rauh & Mangelsdorff

Distribution: Madagascar.

Aloe pruinosa Reynolds 1936

Distribution: South Africa (KwaZulu-Natal).

Aloe pseudorubroviolacea Lavranos & Collen. 2000

Distribution: Saudi Arabia

Aloe pubescens Reynolds 1957

Distribution: Ethiopia.

Aloe pulcherrima M.G.Gilbert & Sebsebe 1997

Distribution: Ethiopia.

Aloe purpurea Lam. 1783
Aloe marginalis DC. 1800 (nom. illeg., Art. 52)
Aloe marginata (Aiton) Willd. 1809 (nom. illeg., Art. 53.1)
Aloe rufocincta Haw. 1819
Dracaena dentata Pers. 1805 (nom. illeg., Art. 52)
Dracaena marginata Aiton 1789 (nom. illeg., Art. 52)
Lomatophyllum aloiflorum (Ker Gawl.) G.Nicholson 1885
Lomatophyllum borbonicum Willd. 1811 (nom. illeg., Art. 52.1)
Lomatophyllum marginatum Hoffmanns. 1824 (nom. nud., Art. 32.1c)
Lomatophyllum purpureum (Lam.) T.Durand & Schinz 1895
Lomatophyllum rufocinctum (Haw.) Salm-Dyck ex Roem. & Schult. 1829
Phylloma aloiflorum Ker Gawl. 1813 (nom. illeg., Art. 52.1)
Phylloma rufocinctum (Haw.) Sweet 1827

Distribution: Mauritius.

Aloe pustuligemma L.E.Newton 1994

Distribution: Kenya.

Aloe ×qaharensis Lavranos & Collen. 2000

Distribution: Saudi Arabia

Aloe rabaiensis Rendle 1895

Distribution: Kenya, Somalia, Tanzania (United Republic of).

Aloe ramosissima Pillans 1937
Aloe dichotoma var. *ramosissima* (Pillans) glen & D.S.Hardy

Distribution: Namibia, South Africa (Northern Cape).

Aloe rauhii Reynolds 1963
Guillauminia rauhii (Reynolds) P.V.Heath 1994

Distribution: Madagascar.

Aloe reitzii Reynolds 1937

Distribution: South Africa.

Aloe reitzii var. **reitzii**

Distribution: South Africa (Mpumalanga).

Aloe reitzii var. **vernalis** D.S.Hardy 1981

Distribution: South Africa (KwaZulu-Natal).

Aloe retrospiciens Reynolds & P.R.O.Bally 1958
Aloe ruspoliana var. *dracaeniformis* A.Berger 1908

Distribution: Somalia.

Aloe reynoldsii Letty 1934

Distribution: South Africa (Eastern Cape).

Aloe rhodesiana Rendle 1911
Aloe eylesii Christian 1936

Distribution: Mozambique, Zimbabwe.

Aloe richardsiae Reynolds 1964

Distribution: Tanzania (United Republic of).

Aloe rigens Reynolds & P.R.O.Bally 1958

Distribution: Yemen, Somalia.

Aloe rigens var. **mortimeri** Lavranos 1967

Distribution: Yemen.

Part II: Aloe

Aloe rigens var. **rigens**

Distribution: Somalia.

Aloe rivae Baker 1898

Distribution: Ethiopia, Kenya.

Aloe rivierei Lavranos & L.E.Newton 1977

Distribution: Yemen.

Aloe rosea (H.Perrier) L.E.Newton & G.D.Rowley 1997
Lomatophyllum roseum H.Perrier 1926

Distribution: Madagascar.

Aloe rubroviolacea Schweinf. 1894

Distribution: Saudi Arabia, Yemen.

Aloe ruffingiana Rauh & Petignat 1999

Distribution: Madagascar

Aloe rugosifolia M.G.Gilbert & Sebsebe 1992
Aloe otallensis var. *elongata* A.Berger 1908

Distribution: Ethiopia, Kenya.

Aloe rupestris Baker 1896
Aloe nitens Baker 1880 (nom. illeg., Art. 53.1)
Aloe pycnacantha MacOwan ms. (nom. nud., Art. 29.1)

Distribution: Mozambique, South Africa (KwaZulu-Natal), Swaziland.

Aloe rupicola Reynolds 1960

Distribution: Angola.

Aloe ruspoliana Baker 1898
Aloe jex-blakeae Christian 1942
Aloe stephaninii Chiov. 1916

Distribution: Ethiopia, Kenya, Somalia.

Aloe sabaea Schweinf. 1894
Aloe gillilandii Reynolds 1962

Distribution: Saudi Arabia, Yemen.

Aloe saundersiae (Reynolds) Reynolds 1947
Aloe minima J.M.Wood 1906 (nom. illeg., Art. 53.1)
Leptaloe saundersiae Reynolds 1936

Distribution: South Africa (KwaZulu-Natal).

Aloe scabrifolia L.E.Newton & Lavranos 1990

Distribution: Kenya.

Aloe schelpei Reynolds 1961

Distribution: Ethiopia.

Aloe schilliana L.E.Newton & G.D.Rowley 1997
Lomatophyllum viviparum H.Perrier 1926

Distribution: Madagascar.

Aloe schoelleri Schweinf. 1894

Distribution: Eritrea.

Aloe schomeri Rauh 1966

Distribution: Madagascar (S.).

Aloe schweinfurthii Baker 1880
Aloe barteri var. *lutea* A.Chev. 1913
Aloe trivialis A.Chev. 1952 (nom. nud., Art. 36.1)

Distribution: Benin, Burkina Faso, Democratic Republic of the Congo (the), Ghana, Mali, Nigeria, Sudan (the), Uganda.

Aloe scobinifolia Reynolds & P.R.O.Bally 1958

Distribution: Somalia.

Aloe scorpioides L.C.Leach 1974

Distribution: Angola.

Part II: Aloe

Aloe secundiflora Engl. 1895

Distribution: Ethiopia, Kenya, Rwanda, Tanzania (United Republic of).

Aloe secundiflora var. **secundiflora**
Aloe engleri A.Berger 1905
Aloe floramaculata Christian 1940
Aloe marsabitensis I.Verd. & Christian 1940

Distribution: Ethiopia, Kenya, Rwanda, Tanzania (United Republic of).

Aloe secundiflora var. **sobolifera** S.Carter 1994

Distribution: Tanzania (United Republic of).

Aloe seretii De Wild. 1921

Distribution: Democratic Republic of the Congo (the).

Aloe serriyensis Lavranos 1965

Distribution: Yemen.

Aloe shadensis Lavranos & Collen. 2000

Distribution: Saudi Arabia.

Aloe sheilae Lavranos 1985

Distribution: Saudi Arabia.

Aloe silicicola H.Perrier 1926

Distribution: Madagascar.

Aloe simii Pole-Evans 1917

Distribution: South Africa (Mpumalanga).

Aloe sinana Reynolds 1957

Distribution: Ethiopia.

Aloe sinkatana Reynolds 1957

Distribution: Sudan (the).

Aloe sladeniana Pole-Evans 1920
Aloe carowii Reynolds 1938

Distribution: Namibia.

Aloe socialis (H.Perrier) L.E.Newton & G.D.Rowley 1997
Lomatophyllum sociale H.Perrier 1926

Distribution: Madagascar.

Aloe somaliensis W.Watson 1899

Distribution: Somalia.

Aloe somaliensis var. **marmorata** Reynolds & P.R.O.Bally 1964

Distribution: Somalia.

Aloe somaliensis var. **somaliensis**

Distribution: Somalia.

Aloe soutpansbergensis I.Verd. 1962

Distribution: South Africa (Northern Prov.).

Aloe speciosa Baker 1880

Distribution: South Africa (Western Cape, Eastern Cape).

Aloe spicata L.f. 1782
Aloe sessiliflora Pole-Evans 1917
Aloe tauri L.C.Leach 1968

Distribution: Mozambique, South Africa (KwaZulu-Natal, Mpumalanga), Swaziland, Zimbabwe.

Aloe splendens Lavranos 1965

Distribution: Yemen.

Aloe squarrosa Baker 1883
Aloe concinna Baker 1898 (nom. illeg., Art. 53.1)
Aloe zanzibarica Milne-Redh. 1947

Distribution: Yemen (Socotra).

Part II: Aloe

Aloe steffanieana Rauh 2000

Distribution: Madagascar

Aloe steudneri Schweinf. 1894

Distribution: Eritrea, Ethiopia.

Aloe striata Haw. 1804

Distribution: Namibia, South Africa.

Aloe striata ssp. **karasbergensis** (Pillans) Glen & D.S.Hardy 1987
Aloe karasbergensis Pillans 1928

Distribution: Namibia, South Africa (Northern Cape).

Aloe striata ssp. **komaggasensis** (Kritzinger & Van Jaarsv.) Glen & D.S.Hardy 1987
Aloe komaggasensis Kritzinger & Van Jaarsv. 1985

Distribution: South Africa (Northern Cape).

Aloe striata ssp. **striata**
Aloe albocincta Haw. 1819
Aloe hanburyana Naudin 1875
Aloe paniculata Jacq. 1809
Aloe rhodocincta hort. ex Baker
Aloe striata var. *oligospila* Baker 1894

Distribution: South Africa (Western Cape, Eastern Cape).

Aloe striatula Haw. 1825

Distribution: Lesotho, South Africa.

Aloe striatula var. **caesia** Reynolds 1936
Aloe striatula forma *conimbricensis* Resende 1943
Aloe striatula forma *haworthii* Resende 1943
Aloe striatula forma *typica* Resende (nom. nud., Art. 24.3)

Distribution: South Africa (Eastern Cape).

Aloe striatula var. **striatula**
Aloe aurantiaca Baker 1892
Aloe cascadensis Kuntze 1898
Aloe macowanii Baker 1880

Distribution: Lesotho, South Africa (Eastern Cape).

Aloe suarezensis H.Perrier 1926

Distribution: Madagascar.

Aloe subacutissima G.D.Rowley 1973
Aloe deltoideodonta var. *intermedia* H.Perrier 1926
Aloe intermedia (H.Perrier) Reynolds 1957 (nom. illeg., Art. 53.1)

Distribution: Madagascar.

Aloe succotrina All. 1773
Aloe perfoliata var. ξ L. 1753
Aloe perfoliata var. *purpurascens* Aiton 1789
Aloe perfoliata var. *succotrina* Aiton 1789
Aloe purpurascens (Aiton) Haw. 1804
Aloe sinuata Thunb. 1794
Aloe soccotorina Schult. & Schult.f. 1829 (nom. nud., Art. 61.1)
Aloe soccotrina Garsault 1767 (nom. nud., Art. 32.8)
Aloe soccotrina var. *purpurascens* Ker Gawl. 1812 (nom. nud., Art. 43.1)
Aloe succotrina Lam. 1783 (nom. illeg., Art. 53.1)
Aloe succotrina var. *saxigena* A.Berger 1908
Aloe vera Mill. 1768 (nom. illeg., Art. 53.1)

Distribution: South Africa (Western Cape).

Aloe suffulta Reynolds 1937
Aloe subfulta hort. (nom. nud., Art. 61.1)

Distribution: Mozambique, South Africa (KwaZulu-Natal).

Aloe suprafoliata Pole-Evans 1916
Aloe suprafoliolata hort. (nom. nud., Art. 61.1)

Distribution: South Africa (KwaZulu-Natal, Mpumalanga), Swaziland.

Aloe suzannae Decary 1921

Distribution: Madagascar.

Aloe swynnertonii Rendle 1911
Aloe chimanimaniensis Christian 1936
Aloe melsetterensis Christian 1938

Distribution: South Africa (Northern Prov.), Zimbabwe.

Aloe tenuior Haw.1825
Aloe tenuior var. *decidua* Reynolds 1936
Aloe tenuior var. *densiflora* Reynolds 1950
Aloe tenuior var. *glaucescens* Zahlbr. 1900
Aloe tenuior var. *rubriflora* Reynolds 1936

Distribution: South Africa (Eastern Cape).

Part II: Aloe

Aloe tewoldei M.G.Gilbert & Sebsebe 1997

Distribution: Ethiopia.

Aloe thompsoniae Groenew. 1936

Distribution: South Africa (Northern Prov.).

Aloe thorncroftii Pole-Evans 1917

Distribution: South Africa (Mpumalanga).

Aloe thraskii Baker 1880
Aloe candelabrum Engl. & Drude 1910 (nom. illeg., Art. 53.1)

Distribution: South Africa (Eastern Cape, KwaZulu-Natal).

Aloe tomentosa Deflers 1889
Aloe tomentosa forma *viridiflora* Lodé 1997 (nom. nud., Art. 34.1b, 36.1)

Distribution: Saudi Arabia, Yemen.

Aloe tormentorii (Marais) L.E.Newton & G.D.Rowley 1997
Lomatophyllum tormentorii Marais 1975

Distribution: Mauritius.

Aloe tororoana Reynolds 1953

Distribution: Uganda.

Aloe torrei I.Verd. & Christian 1946

Distribution: Mozambique.

Aloe trachyticola (H.Perrier) Reynolds 1957
Aloe capitata var. *trachyticola* H.Perrier 1926

Distribution: Madagascar.

Aloe trichosantha A.Berger 1905

Distribution: Eritrea, Ethiopia.

Aloe trichosantha ssp. **longiflora** M.G.Gilbert & Sebsebe 1997

Distribution: Ethiopia.

Aloe trichosantha ssp. **trichosantha**
Aloe percrassa Schweinf. 1894 (nom. illeg., Art. 53.1)
Aloe percrassa var. *albo-picta* Schweinf. 1894 (incorrect name, Art. 11.4)

Distribution: Eritrea, Ethiopia.

Aloe trigonantha L.C.Leach 1971

Distribution: Ethiopia.

Aloe tugenensis L.E.Newton & Lavranos 1990

Distribution: Kenya.

Aloe turkanensis Christian 1942

Distribution: Kenya, Uganda.

Aloe tweediae Christian 1942

Distribution: Kenya, Sudan (the), Uganda.

Aloe ukambensis Reynolds 1956

Distribution: Kenya.

Aloe umfoloziensis Reynolds 1937

Distribution: South Africa (KwaZulu-Natal).

Aloe vacillans Forssk. 1775
Aloe audhalica Lavranos & D.S.Hardy 1965
Aloe dhalensis Lavranos 1965

Distribution: Yemen.

Aloe vallaris L.C.Leach 1974

Distribution: Angola.

Aloe vanbalenii Pillans 1934

Distribution: South Africa (KwaZulu-Natal).

Aloe vandermerwei Reynolds 1950

Distribution: South Africa (Northern Prov.).

Part II: Aloe

Aloe vaombe Decorse & Poiss. 1912

Distribution: Madagascar.

Aloe vaombe var. **poissonii** Decary 1921

Distribution: Madagascar.

Aloe vaombe var. **vaombe**

Distribution: Madagascar.

Aloe vaotsanda Decary 1921

Distribution: Madagascar.

Aloe variegata L. 1753
Aloe ausana Dinter 1931
Aloe punctata Haw. 1804
Aloe variegata var. *haworthii* A.Berger 1908

Distribution: Namibia, South Africa (Northern Cape, Western Cape, Eastern Cape, Free State).

Aloe vera (L.) Burm.f. 1768
Aloe barbadensis Mill. 1768
Aloe barbadensis var. *chinensis* Haw. 1819
Aloe chinensis (Haw.) Baker 1877
Aloe elongata Murray 1789
Aloe flava Pers. 1805
Aloe indica Royle 1839
Aloe lanzae Tod. 1891
Aloe perfoliata var. *barbadensis* (Mill.) Aiton 1789
Aloe perfoliata var. *vera* L. 1753
Aloe vera var. *chinensis* (Haw.) A.Berger 1908
Aloe vera var. *lanzae* (Tod.) A.Berger 1908
Aloe vera var. *littoralis* K.D.Koenig ex Baker 1880
Aloe vera var. *wratislaviensis* Kostecka-Madalska 1953
Aloe vulgaris Lam. 1783

Distribution: Only in cultivation.

Aloe verecunda Pole-Evans 1917

Distribution: South Africa (North-West Prov., Northern Prov., Gauteng, Mpumalanga).

Aloe versicolor Guillaumin 1950

Distribution: Madagascar.

Aloe veseyi Reynolds 1959

Distribution: Tanzania (United Republic of), Zambia.

Aloe viguieri H.Perrier 1927

Distribution: Madagascar.

Aloe viridiflora Reynolds 1937

Distribution: Namibia.

Aloe vituensis Baker 1898

Distribution: Kenya, Sudan (the).

Aloe vogtsii Reynolds 1936

Distribution: South Africa (Northern Prov.).

Aloe volkensii Engl. 1895

Distribution: Kenya, Tanzania (United Republic of), Uganda.

Aloe volkensii ssp. **multicaulis** S.Carter & L.E.Newton 1994

Distribution: Kenya, Tanzania (United Republic of), Uganda.

Aloe volkensii ssp. **volkensii**
Aloe stuhlmannii Baker 1898

Distribution: Kenya, Tanzania (United Republic of).

Aloe vossii Reynolds 1936

Distribution: South Africa (Northern Prov.).

Aloe vryheidensis Groenew. 1937
Aloe dolomitica Groenew. 1938

Distribution: South Africa (KwaZulu-Natal).

Aloe whitcombei Lavranos 1995

Distribution: Oman.

Part II: Aloe

Aloe wildii (Reynolds) Reynolds 1964
Aloe torrei var. *wildii* Reynolds 1961

Distribution: Zimbabwe.

Aloe wilsonii Reynolds 1956

Distribution: Kenya, Uganda.

Aloe wollastonii Rendle 1908
Aloe angiensis De Wild. 1921
Aloe angiensis var. *kitaliensis* Reynolds 1955
Aloe bequaertii De Wild. 1921
Aloe lanuriensis De Wild. 1921
Aloe lateritia var. *kitaliensis* (Reynolds) Reynolds 1966

Distribution: Democratic Republic of the Congo (the), Kenya, Tanzania (United Republic of), Uganda.

Aloe woodii Lavranos & Collen.

Distribution: Saudi Arabia

Aloe wrefordii Reynolds 1956

Distribution: Kenya, Sudan (the), Uganda.

Aloe yavellana Reynolds 1954

Distribution: Ethiopia.

Aloe yemenica J.R.I.Wood 1983

Distribution: Yemen.

Aloe zebrina Baker 1878
Aloe ammophila Reynolds 1936
Aloe angustifolia Groenew. 1938 (nom. illeg., Art. 53.1)
Aloe bamangwatensis Schönland 1904
Aloe laxissima Reynolds 1936
Aloe lugardiana Baker 1901
Aloe platyphylla Baker 1878
Aloe transvaalensis Kuntze 1898
Aloe transvaalensis var. *stenacantha* F.S.Mull. 1940
Aloe vandermerwei Reynolds 1950

Distribution: Angola, Botswana, Malawi, Mozambique, Namibia, South Africa (Gauteng, Mpumalanga, Northern Prov.), Zambia, Zimbabwe.

Aloe zombitsiensis Rauh & M.Teissier 2000

Distribution: Madagascar (SW).

PACHYPODIUM BINOMIALS IN CURRENT USE

PACHYPODIUM BINOMES ACTUELLEMENT EN USAGE

PACHYPODIUM BINOMIALES UTILIZADOS NORMALMENTE

Pachypodium ambongense Poiss. 1924

Distribution: Madagascar (Ambongo).

Pachypodium baronii Costantin & Bois 1907

Distribution: Madagascar (N).

Pachypodium baronii var. **baronii**
Pachypodium baronii var. *erythreum* Poiss. 1924
Pachypodium baronii var. *typicum* Pichon 1949 (nom. nud., Art. 24.3).

Distribution: Madagascar (N).

Pachypodium baronii var. **windsorii** (Poiss.) Pichon 1949
Pachypodium windsorii Poiss. 1917

Distribution: Madagascar (N).

Pachypodium bispinosum (L.f.) A.DC. 1844
Belonites bispinosa (L.f.) E.Mey. 1837
Echites bispinosa L.f. 1781
Pachypodium glabrum G.Don 1838
Pachypodium tuberosum var. *loddigesii* A.DC. 1844

Distribution: South Africa (Eastern Cape).

Pachypodium brevicaule Baker 1887

Distribution: Madagascar (C).

Pachypodium decaryi Poiss. 1916

Distribution: Madagascar (N).

Pachypodium densiflorum Baker 1886
Pachypodium brevicalyx (H.Perrier) Pichon 1949
Pachypodium densiflorum var. *brevicalyx* H.Perrier 1934

Distribution: Madagascar (C).

Part II: Pachypodium

Pachypodium densiflorum var. **densiflorum**

Distribution: Madagascar (C).

Pachypodium geayi Costantin & Bois 1907

Distribution: Madagascar (S).

Pachypodium ×hojnyi Halda 1998

Distribution: only in cultivation.

Pachypodium horombense Poiss. 1924
Pachypodium rosulatum var. *horombense* (Poiss.) G.D.Rowley 1973

Distribution: Madagascar (C and S).

Pachypodium lamerei Drake 1899
Pachypodium champenoisianum Boiteau 1941
Pachypodium lamerei var. *lamerei*
Pachypodium lamerei var. *ramosum* (Costantin & Bois) Pichon 1949
Pachypodium lamerei var. *typicum* Pichon 1949 (nom. nud. Art. 23.4).
Pachypodium menabeum Léandri 1934
Pachypodium ramosum Costantin & Bois 1907
Pachypodium rutenbergianum forma *lamerei* (Drake) Poiss. 1924

Distribution: Madagascar (S).

Pachypodium lealii Welw. 1869
Pachypodium giganteum Engl. 1894

Distribution: Angola, Botswana, Namibia, South Africa, Zimbabwe.

Pachypodium lealii ssp. **lealii**

Distribution: Angola, Botswana, Namibia

Pachypodium lealii ssp. **saundersii** (N.E.Br.) G.D.Rowley 1973
Pachypodium saundersii N.E.Br. 1892

Distribution: South Africa (to N KwaZulu-Natal) to Zimbabwe (S).

Pachypodium namaquanum (Wiley ex Harv.) Welw. 1869
Adenium namaquanum Wiley ex Harv. 1863

Distribution: Namibia (S), South Africa (Northern Cape: Namaqualand).

Pachypodium ×**rauhii** Halda 1997

Distribution: Madagascar (SW).

Pachypodium rosulatum Baker 1882
Pachypodium cactipes K.Schum. 1895
Pachypodium drakei Costantin & Bois 1907

Distribution: Madagascar.

Pachypodium rosulatum forma **bicolor** (Lavranos & Rapan.) G.D.Rowley 1998
Pachypodium bicolor Lavranos & Rapan. 1997

Distribution: Madagascar (W).

Pachypodium rosulatum var. **eburneum** (Lavranos & Rapan.) G.D.Rowley 1998
Pachypodium eburneum Lavranos & Rapan. 1997

Distribution: Madagascar (C).

Pachypodium rosulatum var. **gracilius** H.Perrier 1934
Pachypodium gracilius (H. Perrier) Rapan. 1999

Distribution: Madagascar (Isalo Mts.).

Pachypodium rosulatum var. **inopinatum** (Lavranos) G.D.Rowley 1998
Pachypodium inopinatum Lavranos 1996

Distribution: Madagascar (C).

Pachypodium rosulatum var. **rosulatum**
Pachypodium rosulatum var. *delphinense* H.Perrier 1934
Pachypodium rosulatum var. *drakei* (Costantin & Bois) Markgr. 1976
Pachypodium rosulatum var. *stenanthum* Costantin & Bois 1907
Pachypodium rosulatum var. *typicum* Costantin & Bois 1907 (nom. nud., Art. 24.3)

Distribution: Madagascar.

Pachypodium rosulatum var. **rosulatum** forma **bicolor** (Lavranos & Rapan.) G.D.Rowley 1998

Distribution: Madagascar

Pachypodium rutenbergianum Vatke 1885

Distribution: Madagascar

Part II: Pachypodium

Pachypodium rutenbergianum var. **meridionale** H.Perrier 1934
Pachypodium meridionale (H.Perrier) Pichon 1949.

Distribution: Madagascar (SW)

Pachypodium rutenbergianum var. **rutenbergianum**
Pachypodium rutenbergianum var. *typicum* H.Perrier 1934 (nom. nud., Art. 24.3).

Distribution: Madagascar (NW).

Pachypodium rutenbergianum var. **sofiense** Poiss. 1924
Pachypodium rutenbergianum var. *perrieri* Poiss. 1924
Pachypodium sofiense (Poiss.) H.Perrier 1934

Distribution: Madagascar (NW).

Pachypodium succulentum (L.f.) Sweet 1830
Belonites succulenta (L.f.) E.Mey. 1837
Echites succulenta L.f. 1781
Pachypodium griquense L.Bolus 1932
Pachypodium jasminiflorum L.Bolus 1932.
Pachypodium tomentosum G.Don 1838
Pachypodium tuberosum Lindl. 1830

Distribution: South Africa (Northern Cape, Western Cape, N-wards to Free State).

PART III: COUNTRY CHECKLIST
For the genera:

Aloe and *Pachypodium*

TROISIEME PARTIE: LISTE DES PAYS
Pour les genres:

Aloe et *Pachypodium*

PARTE III: LISTA POR PAISES
Para los géneros:

Aloe y *Pachypodium*

Part III: Country checklist for the genera:
Aloe and *Pachypodium*

Troisième partie: Liste par pays pour les genre:
Aloe et *Pachypodium*

Parte III: Lista por paises para el genero:
Aloe y *Pachypodium*

ONLY IN CULTIVATION

Aloe vera (L.) Burm.f.
Pachypodium ×hojnyi Halda

ANGOLA / ANGOLA (L') / ANGOLA

Aloe andongensis Baker
Aloe andongensis var. **andongensis**
Aloe andongensis var. **repens** L.C.Leach
Aloe angolensis Baker
Aloe bulbicaulis Christian
Aloe catengiana Reynolds
Aloe christianii Reynolds
Aloe esculenta L.C.Leach
Aloe gossweileri Reynolds
Aloe grata Reynolds
Aloe guerrae Reynolds
Aloe hereroensis Engl.
Aloe inamara L.C.Leach
Aloe lepida L.C.Leach
Aloe littoralis Baker
Aloe mendesii Reynolds
Aloe metallica Engl. & Gilg
Aloe milne-redheadii Christian
Aloe nuttii Baker
Aloe paedogona A.Berger
Aloe palmiformis Baker
Aloe procera L.C.Leach
Aloe rupicola Reynolds
Aloe scorpioides L.C.Leach
Aloe vallaris L.C.Leach
Aloe zebrina Baker
Pachypodium lealii Welw.
Pachypodium lealii ssp. **lealii**

BENIN / BÉNIN (LE) / BENIN

Aloe buettneri A.Berger
Aloe macrocarpa Tod.
Aloe schweinfurthii Baker

Part III: Country Checklist

BOTSWANA / BOTSWANA (LE) / BOTSWANA

Aloe cryptopoda Baker
Aloe esculenta L.C.Leach
Aloe globuligemma Pole-Evans
Aloe grandidentata Salm-Dyck
Aloe greatheadii Schönland
Aloe greatheadii var. **greatheadii**
Aloe littoralis Baker
Aloe marlothii A.Berger
Aloe marlothii var. **marlothii**
Aloe zebrina Baker
Pachypodium lealii Welw.
Pachypodium lealii ssp. **lealii**

BURKINA FASO / BURKINA FASO (LE) / BURKINA FASO

Aloe schweinfurthii Baker

BURUNDI / BURUNDI (LE) / BURUNDI

Aloe myriacantha (Haw.) Schult. & Schult.f.

CAMEROON / CAMEROUN (LE) / CAMERÚN (EL)

Aloe macrocarpa Tod.

COMOROS (THE) / COMORES (LES) / COMORAS (LAS)

Aloe mayottensis A.Berger

DEMOCRATIC REPUBLIC OF THE CONGO (THE) / RÉPUBLIQUE DÉMOCRATIQUE DU CONGA (LA) / REPÚBLICA DEMOCRÁTICA DEL CONGO (LA)

Aloe bulbicaulis Christian
Aloe chabaudii Schönland
Aloe chabaudii var. **chabaudii**
Aloe christianii Reynolds
Aloe dawei A.Berger
Aloe greatheadii Schönland
Aloe greatheadii var. **greatheadii**
Aloe hendrickxii Reynolds
Aloe myriacantha (Haw.) Schult. & Schult.f.
Aloe mzimbana Christian
Aloe nuttii Baker
Aloe patersonii B.Mathew
Aloe schweinfurthii Baker
Aloe seretii De Wild.
Aloe wollastonii Rendle

DJIBOUTI / DJIBOUTI / DJIBOUTI

Aloe macrocarpa Tod.

DJIBOUTI (Continued)

Aloe mcloughlinii Christian

ERITREA / ERYTHRÉE (L') / ERITREA

Aloe camperi Schweinf.
Aloe elegans Tod.
Aloe eumassawana S.Carter & al.
Aloe macrocarpa Tod.
Aloe percrassa Tod.
Aloe schoelleri Schweinf.
Aloe steudneri Schweinf.
Aloe trichosantha A.Berger
Aloe trichosantha ssp. **trichosantha**

ETHIOPIA / ETHIOPIE (L') / ETIOPÍA

Aloe adigratana Reynolds
Aloe ankoberensis M.G.Gilbert & Sebsebe
Aloe bertemariae Sebsebe & Dioli
Aloe calidophila Reynolds
Aloe camperi Schweinf.
Aloe citrina S.Carter & Brandham
Aloe debrana Christian
Aloe elegans Tod.
Aloe ellenbeckii A.Berger
Aloe frissii Sebsebe & M.G.Gilbert
Aloe gilbertii T.Reynolds ex Sebsebe & Brandham
Aloe gilbertii ssp. **gilbertii**
Aloe gilbertii ssp. **megalacanthoides** M.G.Gilbert & Sebsebe
Aloe harlana Reynolds
Aloe jacksonii Reynolds
Aloe kefaensis M.G.Gilbert & Sebsebe
Aloe lateritia Engl.
Aloe lateritia var. **graminicola** (Reynolds) S.Carter
Aloe macrocarpa Tod.
Aloe mcloughlinii Christian
Aloe megalacantha Baker
Aloe megalacantha ssp. **alticola** M.G.Gilbert & Sebsebe
Aloe megalacantha ssp. **megalacantha**
Aloe monticola Reynolds
Aloe otallensis Baker
Aloe parvidens M.G.Gilbert & Sebsebe
Aloe percrassa Tod.
Aloe pirottae A.Berger
Aloe pubescens Reynolds
Aloe pulcherrima M.G.Gilbert & Sebsebe
Aloe rivae Baker
Aloe rugosifolia M.G.Gilbert & Sebsebe
Aloe ruspoliana Baker
Aloe schelpei Reynolds
Aloe secundiflora Engl.

Part III: Country Checklist

ETHIOPIA (Continued)

Aloe secundiflora var. **secundiflora**
Aloe sinana Reynolds
Aloe steudneri Schweinf.
Aloe tewoldei M.G.Gilbert & Sebsebe
Aloe trichosantha A.Berger
Aloe trichosantha ssp. **longiflora** M.G.Gilbert & Sebsebe
Aloe trichosantha ssp. **trichosantha**
Aloe trigonantha L.C.Leach
Aloe yavellana Reynolds

GHANA / GHANA (LE) / GHANA

Aloe buettneri A.Berger
Aloe ×keayi Reynolds (pro sp.)
Aloe macrocarpa Tod.
Aloe schweinfurthii Baker

KENYA / KENYA (LA) / KENYA

Aloe aageodonta L.E.Newton
Aloe amicorum L.E.Newton
Aloe amudatensis Reynolds
Aloe archeri Lavranos
Aloe ballyi Reynolds
Aloe calidophila Reynolds
Aloe cheranganiensis S.Carter & Brandham
Aloe chrysostachys Lavranos & L.E.Newton
Aloe citrina S.Carter & Brandham
Aloe classenii Reynolds
Aloe confusa Engl.
Aloe dawei A.Berger
Aloe deserti A.Berger
Aloe elata S.Carter & L.E.Newton
Aloe elgonica Bullock
Aloe ellenbeckii A.Berger
Aloe erensii Christian
Aloe fibrosa Lavranos & L.E.Newton
Aloe francombei L.E.Newton
Aloe juvenna Brandham & S.Carter
Aloe kedongensis Reynolds
Aloe ketabrowniorum L.E.Newton
Aloe kilifiensis Christian
Aloe kulalensis L.E.Newton & Beentje
Aloe lateritia Engl.
Aloe lateritia var. **graminicola** (Reynolds) S.Carter
Aloe lateritia var. **lateritia**
Aloe lensayuensis Lavranos & L.E.Newton
Aloe lolwensis L.E. Newton
Aloe macrosiphon Baker
Aloe microdonta Chiov.
Aloe morijensis S.Carter & Brandham

KENYA (Continued)

Aloe multicolor L.E.Newton
Aloe murina L.E.Newton
Aloe myriacantha (Haw.) Schult. & Schult.f.
Aloe ngongensis Christian
Aloe nyeriensis Christian ex I.Verd.
Aloe parvidens M.G.Gilbert & Sebsebe
Aloe penduliflora Baker
Aloe pirottae A.Berger
Aloe powysiorum L.E.Newton & Beentje
Aloe pustuligemma L.E.Newton
Aloe rabaiensis Rendle
Aloe rivae Baker
Aloe rugosifolia M.G.Gilbert & Sebsebe
Aloe ruspoliana Baker
Aloe scabrifolia L.E.Newton & Lavranos
Aloe secundiflora Engl.
Aloe secundiflora var. secundiflora
Aloe tugenensis L.E.Newton & Lavranos
Aloe turkanensis Christian
Aloe tweediae Christian
Aloe ukambensis Reynolds
Aloe vituensis Baker
Aloe volkensii Engl.
Aloe volkensii ssp. multicaulis S.Carter & L.E.Newton
Aloe volkensii ssp. volkensii
Aloe wilsonii Reynolds
Aloe wollastonii Rendle
Aloe wrefordii Reynolds

LESOTHO / LESOTHO (LE) / LESOTHO

Aloe aristata Haw.
Aloe broomii Schönland
Aloe broomii var. broomii
Aloe ecklonis Salm-Dyck
Aloe ferox Mill.
Aloe maculata All.
Aloe polyphylla Schönland ex Pillans
Aloe pratensis Baker
Aloe striatula Haw.
Aloe striatula var. striatula

MADAGASCAR / MADAGASCAR / MADAGASCAR

Aloe acutissima H.Perrier
Aloe acutissima var. acutissima
Aloe acutissima var. antanimorensis Reynolds
Aloe albiflora Guillaumin
Aloe alfredii Rauh
Aloe andringitrensis H.Perrier
Aloe anivoranoensis (Rauh & Hebding) L.E.Newton & G.D.Rowley

MADAGASCAR (Continued)

Aloe ankaranensis Rauh & Mangelsdorff
Aloe antandroi (Decary) H.Perrier
Aloe antsingyensis (Léandri) L.E.Newton & G.D.Rowley
Aloe bakeri Scott-Elliot
Aloe belavenokensis (Rauh & Gerold) L.E.Newton & G.D.Rowley
Aloe bellatula Reynolds
Aloe berevoana Lavranos
Aloe bernadettae J.-B.Castillon
Aloe betsileensis H.Perrier
Aloe boiteaui Guillaumin
Aloe bosseri J.-B.Castillon
Aloe buchlohii Rauh
Aloe bulbillifera H.Perrier
Aloe bulbillifera var. **bulbillifera**
Aloe bulbillifera var. **paulianae** Reynolds
Aloe calcairophila Reynolds
Aloe capitata Baker
Aloe capitata var. **capitata**
Aloe capitata var. **cipolinicola** H.Perrier
Aloe capitata var. **gneissicola** H.Perrier
Aloe capitata var. **quartziticola** H.Perrier
Aloe capitata var. **silvicola** H.Perrier
Aloe capmanambatoensis Rauh & Gerold
Aloe citrea (Guillaumin) L.E.Newton & G.D.Rowley
Aloe compressa H.Perrier
Aloe compressa var. **compressa**
Aloe compressa var. **paucituberculata** Lavranos
Aloe compressa var. **rugosquamosa** H.Perrier
Aloe compressa var. **schistophila** H.Perrier
Aloe conifera H.Perrier
Aloe cremersii Lavranos
Aloe cryptoflora Reynolds
Aloe cyrtophylla Lavranos
Aloe decorsei H.Perrier
Aloe delphinensis Rauh
Aloe deltoideodonta Baker
Aloe deltoideodonta var. **brevifolia** H.Perrier
Aloe deltoideodonta var. **candicans** H.Perrier
Aloe deltoideodonta var. **deltoideodonta**
Aloe descoingsii Reynolds
Aloe descoingsii ssp. **augustina** Lavranos
Aloe descoingsii ssp. **descoingsii**
Aloe divaricata A.Berger
Aloe divaricata var. **divaricata**
Aloe divaricata var. **rosea** (Decary) Reynolds
Aloe ericetorum Bosser
Aloe erythrophylla Bosser
Aloe fievetii Reynolds
Aloe fleuretteana Rauh & Gerold
Aloe fragilis Lavranos & Röösli
Aloe guillaumetii Cremers

MADAGASCAR (Continued)

Aloe haworthioides Baker
Aloe haworthioides var. **aurantiaca** H.Perrier.
Aloe haworthioides var. **haworthioides**
Aloe helenae Danguy
Aloe humbertii H.Perrier
Aloe ibitiensis H.Perrier
Aloe imalotensis Reynolds
Aloe ×imerinensis Bosser.
Aloe isaloensis H.Perrier
Aloe itremensis Reynolds
Aloe laeta A.Berger
Aloe laeta var. **laeta**
Aloe laeta var. **maniaensis** H.Perrier
Aloe leandrii Bosser
Aloe lucile-allorgeae Rauh
Aloe macroclada Baker
Aloe madecassa H.Perrier
Aloe madecassa var. **lutea** Guillaumin
Aloe madecassa var. **madecassa**
Aloe megalocarpa Lavranos
Aloe millotii Reynolds
Aloe namorokaensis (Rauh) L.E.Newton & G.D.Rowley
Aloe occidentalis (H.Perrier) L.E.Newton & G.D.Rowley
Aloe oligophylla Baker
Aloe orientalis (H.Perrier) L.E.Newton & G.D.Rowley
Aloe parallelifolia H.Perrier
Aloe parvula A.Berger
Aloe perrieri Reynolds
Aloe peyrierasii Cremers
Aloe propagulifera (Rauh & Razaf.) L.E.Newton & G.D.Rowley
Aloe prostrata (H.Perrier) L.E.Newton & G.D.Rowley
Aloe prostrata ssp. **pallida** Rauh & Mangelsdorff
Aloe rauhii Reynolds
Aloe rosea (H.Perrier) L.E.Newton & G.D.Rowley
Aloe ruffingiana Rauh & Petignat
Aloe schilliana L.E.Newton & G.D.Rowley
Aloe schomeri Rauh
Aloe silicicola H.Perrier
Aloe socialis (H.Perrier) L.E.Newton & G.D.Rowley.
Aloe steffanieana Rauh
Aloe suarezensis H.Perrier
Aloe subacutissima G.D.Rowley
Aloe suzannae Decary
Aloe trachyticola (H.Perrier) Reynolds
Aloe vaombe Decorse & Poiss.
Aloe vaombe var. **poissonii** Decary
Aloe vaombe var. **vaombe**
Aloe vaotsanda Decary
Aloe versicolor Guillaumin
Aloe viguieri H.Perrier
Aloe zombitsiensis Rauh & M. Teissier

MADAGASCAR (Continued)

Pachypodium ambongense Poiss.
Pachypodium baronii Costantin & Bois
Pachypodium baronii var. baronii
Pachypodium baronii var. windsorii (Poiss.) Pichon
Pachypodium brevicaule Baker
Pachypodium decaryi Poiss.
Pachypodium densiflorum Baker
Pachypodium densiflorum var. densiflorum
Pachypodium geayi Costantin & Bois
Pachypodium horombense Poiss.
Pachypodium lamerei Drake
Pachypodium ×rauhii Halda
Pachypodium rosulatum Baker
Pachypodium rosulatum forma bicolor (Lavranos & Rapan.) G.D.Rowley
Pachypodium rosulatum var. eburneum (Lavranos & Rapan.) G.D.Rowley
Pachypodium rosulatum var. gracilius H.Perrier
Pachypodium rosulatum var. inopinatum (Lavranos) G.D.Rowley
Pachypodium rosulatum var. rosulatum
Pachypodium rosulatum var. rosulatum forma bicolor (Lavranos & Rapan.) G.D.Rowley
Pachypodium rutenbergianum Vatke
Pachypodium rutenbergianum var. meridionale H.Perrier
Pachypodium rutenbergianum var. rutenbergianum
Pachypodium rutenbergianum var. sofiense Poiss.

MALAWI / MALAWI (LE) / MALAWI

Aloe arborescens Mill.
Aloe buchananii Baker
Aloe bulbicaulis Christian
Aloe cameronii Hemsl.
Aloe cameronii var. cameronii
Aloe cameronii var. dedzana Reynolds
Aloe chabaudii Schönland
Aloe chabaudii var. chabaudii
Aloe chabaudii var. mlanjeana Christian
Aloe christianii Reynolds
Aloe cryptopoda Baker
Aloe duckeri Christian
Aloe excelsa A.Berger
Aloe excelsa var. breviflora L.C.Leach
Aloe excelsa var. excelsa
Aloe greatheadii Schönland
Aloe greatheadii var. greatheadii
Aloe mawii Christian
Aloe menyharthii Baker
Aloe menyharthii ssp. menyharthii
Aloe myriacantha (Haw.) Schult. & Schult.f.
Aloe mzimbana Christian
Aloe nuttii Baker
Aloe zebrina Baker

MALI / MALI (LE) / MALI

Aloe buettneri A.Berger
Aloe macrocarpa Tod.
Aloe schweinfurthii Baker

MAURITIUS / MAURICE / MAURICIO

Aloe purpurea Lam.
Aloe tormentorii (Marais) L.E.Newton & G.D.Rowley

MOZAMBIQUE / MOZAMBIQUE (LE) / MOZAMBIQUE

Aloe arborescens Mill.
Aloe ballii Reynolds
Aloe ballii var. ballii
Aloe ballii var. makurupiniensis Ellert
Aloe barberae Dyer
Aloe cameronii Hemsl.
Aloe cameronii var. cameronii
Aloe cameronii var. dedzana Reynolds
Aloe cannellii L.C.Leach
Aloe chabaudii Schönland
Aloe chabaudii var. chabaudii
Aloe chabaudii var. verekeri Christian
Aloe christianii Reynolds
Aloe cryptopoda Baker
Aloe decurva Reynolds
Aloe excelsa A.Berger
Aloe excelsa var. breviflora L.C.Leach
Aloe excelsa var. excelsa
Aloe greatheadii Schönland
Aloe greatheadii var. greatheadii
Aloe hazeliana Reynolds
Aloe littoralis Baker
Aloe marlothii A.Berger
Aloe marlothii ssp. orientalis Glen & D.S.Hardy
Aloe marlothii var. marlothii
Aloe mawii Christian
Aloe menyharthii Baker
Aloe menyharthii ssp. ensifolia S.Carter
Aloe menyharthii ssp. menyharthii
Aloe munchii Christian
Aloe parvibracteata Schönland
Aloe plowesii Reynolds
Aloe rhodesiana Rendle
Aloe rupestris Baker
Aloe spicata L.f.
Aloe suffulta Reynolds
Aloe torrei I.Verd. & Christian
Aloe zebrina Baker

Part III: Country Checklist

NAMIBIA / NAMIBIE (LA) / NAMIBIA

Aloe argenticauda Merxm. & Giess
Aloe asperifolia A.Berger
Aloe claviflora Burch.
Aloe corallina I.Verd.
Aloe dewinteri Giess
Aloe dichotoma Masson
Aloe dinteri A.Berger
Aloe esculenta L.C.Leach
Aloe gariepensis Pillans
Aloe hereroensis Engl.
Aloe littoralis Baker
Aloe melanacantha A.Berger
Aloe melanacantha var. **erinacea** (D.S.Hardy) G.D.Rowley
Aloe melanacantha var. **melanacantha**
Aloe mendesii Reynolds
Aloe meyeri Van Jaarsv.
Aloe namibensis Giess
Aloe pachygaster Dinter
Aloe paedogona A.Berger
Aloe pearsonii Schönland
Aloe pillansii L.Guthrie
Aloe ramosissima Pillans
Aloe sladeniana Pole-Evans
Aloe striata Haw.
Aloe striata ssp. **karasbergensis** (Pillans) Glen & D.S.Hardy
Aloe variegata L.
Aloe viridiflora Reynolds
Aloe zebrina Baker
Pachypodium lealii Welw.
Pachypodium lealii ssp. **lealii**
Pachypodium namaquanum (Wiley ex Harv.) Welw.

NIGERIA / NIGÉRIA (LE) / NIGERIA

Aloe buettneri A.Berger
Aloe macrocarpa Tod.

OMAN / OMAN (L') / OMÁN

Aloe collenetteae Lavranos
Aloe dhufarensis Lavranos
Aloe luntii Baker
Aloe whitcombei Lavranos

RÉUNION (French) / RÉUNION (La Francais) / RÉUNION (La francesa)

Aloe macra Haw.

RODRIGUES (MAURITIUS) / RODRIGUES (MAURICE) / RODRIGUES (MAURICIO)

Aloe lomatophylloides Balf.f.

RWANDA / RWANDA (LE) / RWANDA

Aloe dawei A.Berger
Aloe macrosiphon Baker
Aloe myriacantha (Haw.) Schult. & Schult.f.
Aloe secundiflora Engl.
Aloe secundiflora var. **secundiflora**

SAUDI ARABIA / ARABIE SAOUDITE (L') / ARABIA SAUDITA

Aloe armatissima Lavranos & Collen.
Aloe brunneodentata Lavranos & Collen.
Aloe castellorum J.R.I.Wood
Aloe cephalophora Lavranos & Collen.
Aloe edentata Lavranos & Collen.
Aloe hijazensis Lavranos & Collen.
Aloe officinalis Forssk.
Aloe parvicapsula Lavranos & Collen.
Aloe parvicoma Lavranos & Collen.
Aloe porphyrostachys Lavranos & Collen.
Aloe pseudorubroviolacea Lavranos & Collen.
Aloe ×qaharensis Lavranos & Collen.
Aloe rubroviolacea Schweinf.
Aloe sabaea Schweinf.
Aloe shadensis Lavranos & Collen.
Aloe sheilae Lavranos
Aloe tomentosa Deflers
Aloe woodii Lavranos & Collen.

SEYCHELLES / SEYCHELLES (LES) / SEYCHELLES

Aloe aldabrensis (Marais) L.E.Newton & G.D.Rowley

SOMALIA / SOMALIE (LA) / SOMALIA

Aloe albovestita S.Carter & Brandham
Aloe ambigens Chiov.
Aloe bargalensis Lavranos
Aloe bella G.D.Rowley
Aloe breviscapa Reynolds & P.R.O.Bally
Aloe brunneostriata Lavranos & S.Carter
Aloe citrina S.Carter & Brandham
Aloe cremnophila Reynolds & P.R.O.Bally
Aloe ellenbeckii A.Berger
Aloe eminens Reynolds & P.R.O.Bally
Aloe gillettii S.Carter
Aloe glabrescens (Reynolds & P.R.O.Bally) S.Carter & Brandham
Aloe gracilicaulis Reynolds & P.R.O.Bally
Aloe grisea S.Carter & Brandham
Aloe heliderana Lavranos
Aloe hemmingii Reynolds & P.R.O.Bally
Aloe heybensis Lavranos 1999
Aloe hildebrandtii Baker
Aloe jucunda Reynolds

SOMALIA (Continued)

Aloe lindenii Lavranos
Aloe luntii Baker
Aloe medishiana Reynolds & P.R.O.Bally
Aloe megalacantha Baker
Aloe megalacantha ssp. **megalacantha**
Aloe microdonta Chiov.
Aloe molederana Lavranos & Glen
Aloe parvidens M.G.Gilbert & Sebsebe
Aloe peckii P.R.O.Bally & I.Verd.
Aloe rabaiensis Rendle
Aloe retrospiciens Reynolds & P.R.O.Bally
Aloe rigens Reynolds & P.R.O.Bally
Aloe rigens var. **rigens**
Aloe ruspoliana Baker
Aloe scobinifolia Reynolds & P.R.O.Bally
Aloe somaliensis W.Watson.
Aloe somaliensis var. **marmorata** Reynolds & P.R.O.Bally
Aloe somaliensis var. **somaliensis**

SOUTH AFRICA / AFRIQUE DU SUD (L') / SUDÁFRICA

Aloe aculeata Pole-Evans
Aloe affinis A.Berger
Aloe africana Mill.
Aloe albida (Stapf) Reynolds
Aloe alooides (Bolus) Druten
Aloe angelica Pole-Evans
Aloe arborescens Mill.
Aloe arenicola Reynolds
Aloe aristata Haw.
Aloe barberae Dyer
Aloe barbertoniae Pole-Evans
Aloe bowiea Schult. & Schult.f.
Aloe boylei Baker
Aloe boylei ssp. **boylei**
Aloe boylei ssp. **major** Hilliard & B.L.Burtt
Aloe branddraaiensis Groenew.
Aloe brevifolia Mill.
Aloe brevifolia var. **brevifolia**
Aloe brevifolia var. **depressa** (Haw.) Baker
Aloe brevifolia var. **postgenita** (M.Roem. & Schult.) Baker
Aloe broomii Schönland
Aloe broomii var. **broomii**
Aloe broomii var. **tarkaensis** Reynolds
Aloe buhrii Lavranos
Aloe burgersfortensis Reynolds
Aloe castanea Schönland
Aloe chabaudii Schönland
Aloe chabaudii var. **chabaudii**
Aloe chlorantha Lavranos
Aloe chortolirioides A.Berger

SOUTH AFRICA (Continued)

Aloe chortolirioides var. **chortolirioides**
Aloe chortolirioides var. **woolliana** (Pole-Evans) Glen & D.S.Hardy
Aloe ciliaris Haw.
Aloe ciliaris var. **ciliaris**
Aloe ciliaris var. **redacta** S.Carter
Aloe ciliaris var. **tidmarshii** Schönland
Aloe claviflora Burch.
Aloe commixta A.Berger
Aloe comosa Marloth & A.Berger
Aloe comptonii Reynolds
Aloe cooperi Baker
Aloe cooperi ssp. **cooperi**
Aloe cooperi ssp. **pulchra** Glen & D.S.Hardy
Aloe cryptopoda Baker
Aloe dabenorisana Van Jaarsv.
Aloe dewetii Reynolds
Aloe dichotoma Masson
Aloe distans Haw.
Aloe dominella Reynolds
Aloe dyeri Schönland
Aloe ecklonis Salm-Dyck
Aloe excelsa A.Berger
Aloe excelsa var. **excelsa**
Aloe falcata Baker
Aloe ferox Mill.
Aloe fosteri Pillans
Aloe fouriei D.S.Hardy & Glen
Aloe framesii L.Bolus
Aloe gariepensis Pillans
Aloe gerstneri Reynolds
Aloe glauca Mill.
Aloe glauca var. **glauca**
Aloe glauca var. **spinosior** Haw.
Aloe globuligemma Pole-Evans
Aloe gracilis Haw.
Aloe gracilis var. **decumbens** Reynolds
Aloe gracilis var. **gracilis**
Aloe grandidentata Salm-Dyck
Aloe greatheadii Schönland
Aloe greatheadii var. **davyana** (Schönland) Glen & D.S.Hardy
Aloe greatheadii var. **greatheadii**
Aloe greenii Baker
Aloe haemanthifolia A.Berger & Marloth
Aloe hardyi Glen
Aloe hereroensis Engl.
Aloe hlangapies Groenew.
Aloe humilis (L.) Mill.
Aloe immaculata Pillans
Aloe inconspicua Plowes
Aloe integra Reynolds
Aloe khamiesensis Pillans

SOUTH AFRICA (Continued)

Aloe kniphofioides Baker
Aloe krapohliana Marloth
Aloe krapohliana var. **dumoulinii** Lavranos
Aloe krapohliana var. **krapohliana**
Aloe kraussii Baker
Aloe lettyae Reynolds
Aloe linearifolia A.Berger
Aloe lineata (Aiton) Haw.
Aloe lineata var. **lineata**
Aloe lineata var. **muirii** (Marloth) Reynolds
Aloe littoralis Baker
Aloe longistyla Baker
Aloe lutescens Groenew.
Aloe maculata All.
Aloe marlothii A.Berger
Aloe marlothii ssp. **orientalis** Glen & D.S.Hardy
Aloe marlothii var. **bicolor** Reynolds
Aloe marlothii var. **marlothii**
Aloe melanacantha A.Berger
Aloe melanacantha var. **melanacantha**
Aloe meyeri Van Jaarsv.
Aloe micracantha Haw.
Aloe microstigma Salm-Dyck
Aloe minima Baker
Aloe minima var. **blyderivierensis** (Groenew.) Reynolds
Aloe minima var. **minima**
Aloe mitriformis Mill.
Aloe modesta Reynolds
Aloe monotropa I.Verd.
Aloe mudenensis Reynolds
Aloe mutabilis Pillans
Aloe myriacantha (Haw.) Schult. & Schult.f.
Aloe nubigena Groenew.
Aloe parvibracteata Schönland
Aloe pearsonii Schönland
Aloe peglerae Schönland
Aloe petricola Pole-Evans
Aloe petrophila Pillans
Aloe pictifolia D.S.Hardy
Aloe pillansii L.Guthrie
Aloe plicatilis (L.) Mill.
Aloe pluridens Haw.
Aloe pratensis Baker
Aloe pretoriensis Pole-Evans
Aloe prinslooi I.Verd. & D.S.Hardy
Aloe pruinosa Reynolds
Aloe ramosissima Pillans
Aloe reitzii Reynolds
Aloe reitzii var. **reitzii**
Aloe reitzii var. **vernalis** D.S.Hardy
Aloe reynoldsii Letty

SOUTH AFRICA (Continued)

Aloe rupestris Baker
Aloe saundersiae (Reynolds) Reynolds
Aloe simii Pole-Evans
Aloe soutpansbergensis I.Verd.
Aloe speciosa Baker
Aloe spicata L.f.
Aloe striata Haw.
Aloe striata ssp. **karasbergensis** (Pillans) Glen & D.S.Hardy
Aloe striata ssp. **komaggasensis** (Kritzinger & Van Jaarsv.) Glen & D.S.Hardy
Aloe striata ssp. **striata**
Aloe striatula Haw.
Aloe striatula var. **caesia** Reynolds
Aloe striatula var. **striatula**
Aloe succotrina All.
Aloe suffulta Reynolds
Aloe suprafoliata Pole-Evans
Aloe swynnertonii Rendle
Aloe tenuior Haw.
Aloe thompsoniae Groenew.
Aloe thorncroftii Pole-Evans
Aloe thraskii Baker
Aloe umfoloziensis Reynolds
Aloe vanbalenii Pillans
Aloe vandermerwei Reynolds 1950
Aloe variegata L.
Aloe verecunda Pole-Evans
Aloe vogtsii Reynolds
Aloe vossii Reynolds
Aloe vryheidensis Groenew.
Aloe zebrina Baker
Pachypodium bispinosum (L.f.) A.DC.
Pachypodium lealii Welw.
Pachypodium namaquanum (Wiley ex Harv.) Welw.
Pachypodium lealii ssp. **saundersii** (N.E.Br.) G.D.Rowley
Pachypodium succulentum (L.f.) Sweet

SUDAN (THE) / SOUDAN (LE) / SUDÁN

Aloe canarina S.Carter
Aloe crassipes Baker
Aloe diolii L.E.Newton
Aloe erensii Christian
Aloe labworana (Reynolds) S.Carter
Aloe macleayi Reynolds
Aloe macrocarpa Tod.
Aloe sinkatana Reynolds
Aloe tweediae Christian
Aloe vituensis Baker
Aloe wrefordii Reynolds

SWAZILAND / SWAZILAND (LE) / SWAZILANDIA

Aloe barberae Dyer 1874
Aloe chabaudii Schönland
Aloe chabaudii var. **chabaudii**
Aloe chortolirioides A.Berger
Aloe chortolirioides var. **chortolirioides**
Aloe chortolirioides var. **woolliana** (Pole-Evans) Glen & D.S.Hardy
Aloe cooperi Baker
Aloe cooperi ssp. **cooperi**
Aloe cooperi ssp. **pulchra** Glen & D.S.Hardy
Aloe cryptopoda Baker
Aloe dewetii Reynolds
Aloe ecklonis Salm-Dyck
Aloe greatheadii Schönland
Aloe greatheadii var. **davyana** (Schönland) Glen & D.S.Hardy
Aloe integra Reynolds
Aloe keithii Reynolds
Aloe maculata All.
Aloe marlothii A.Berger
Aloe marlothii var. **marlothii**
Aloe minima Baker
Aloe minima var. **minima**
Aloe parvibracteata Schönland
Aloe pretoriensis Pole-Evans
Aloe rupestris Baker
Aloe spicata L.f.
Aloe suprafoliata Pole-Evans

TANZANIA (UNITED REPUBLIC OF) / TANZANIE (RÉPUBLIQUE-UNIE DE) (LA) / TANZANÍA (REPÚBLICA UNIDA DE) (LA)

Aloe babatiensis Christian & I.Verd.
Aloe ballyi Reynolds
Aloe bicomitum L.C.Leach
Aloe boscawenii Christian
Aloe brachystachys Baker
Aloe brandhamii S.Carter
Aloe bukobana Reynolds
Aloe bulbicaulis Christian
Aloe bullockii Reynolds
Aloe bussei A.Berger
Aloe chabaudii Schönland
Aloe chabaudii var. **chabaudii**
Aloe christianii Reynolds
Aloe confusa Engl.
Aloe congdonii S.Carter
Aloe deserti A.Berger
Aloe dorotheae A.Berger
Aloe duckeri Christian
Aloe elata S.Carter & L.E.Newton
Aloe fibrosa Lavranos & L.E.Newton
Aloe fimbrialis S.Carter

TANZANIA (UNITED REPUBLIC OF) (Continued)

Aloe flexilifolia Christian
Aloe lateritia Engl.
Aloe lateritia var. **lateritia**
Aloe leachii Reynolds
Aloe leedalii S.Carter
Aloe leptosiphon A.Berger
Aloe macrosiphon Baker
Aloe massawana Reynolds
Aloe mawii Christian
Aloe morijensis S.Carter & Brandham
Aloe myriacantha (Haw.) Schult. & Schult.f.
Aloe mzimbana Christian
Aloe ngongensis Christian
Aloe nuttii Baker
Aloe parvidens M.G.Gilbert & Sebsebe
Aloe pembana L.E.Newton
Aloe rabaiensis Rendle
Aloe richardsiae Reynolds
Aloe secundiflora Engl.
Aloe secundiflora var. **secundiflora**
Aloe secundiflora var. **sobolifera** S.Carter
Aloe veseyi Reynolds
Aloe volkensii Engl.
Aloe volkensii ssp. **multicaulis** S.Carter & L.E.Newton
Aloe volkensii ssp. **volkensii**
Aloe wollastonii Rendle

TOGO / TOGO (LE) / TOGO (EL)

Aloe buettneri A.Berger

UGANDA / OUGANDA (L') / UGANDA

Aloe amudatensis Reynolds
Aloe canarina S.Carter
Aloe cheranganiensis S.Carter & Brandham
Aloe dawei A.Berger
Aloe labworana (Reynolds) S.Carter
Aloe macrosiphon Baker
Aloe mubendiensis Christian
Aloe myriacantha (Haw.) Schult. & Schult.f.
Aloe tororoana Reynolds
Aloe turkanensis Christian
Aloe tweediae Christian
Aloe volkensii Engl.
Aloe volkensii ssp. **multicaulis** S.Carter & L.E.Newton
Aloe wilsonii Reynolds
Aloe wollastonii Rendle
Aloe wrefordii Reynolds

YEMEN / YÉMEN / YEMEN (EL)

Aloe abyssicola Lavranos & Bilaidi
Aloe ahmarensis Favell, M.B.Mill. & Al-Gifri
Aloe castellorum J.R.I.Wood.
Aloe doei Lavranos
Aloe doei var. **doei**
Aloe doei var. **lavranosii** Marn.-Lap.
Aloe eremophila Lavranos
Aloe fleurentiniorum Lavranos & L.E.Newton
Aloe forbesii Balf.f.
Aloe fulleri Lavranos
Aloe inermis Forssk.
Aloe lavranosii Reynolds
Aloe luntii Baker
Aloe menachensis (Schweinf.) Blatt.
Aloe niebuhriana Lavranos
Aloe officinalis Forssk.
Aloe pendens Forssk.
Aloe perryi Baker
Aloe rigens Reynolds & P.R.O.Bally
Aloe rigens var. **mortimeri** Lavranos
Aloe rivierei Lavranos & L.E.Newton
Aloe rubroviolacea Schweinf.
Aloe sabaea Schweinf.
Aloe serriyensis Lavranos
Aloe splendens Lavranos
Aloe squarrosa Baker
Aloe tomentosa Deflers
Aloe vacillans Forssk.
Aloe yemenica J.R.I.Wood

ZAMBIA / ZAMBIE (LA) / ZAMBIA

Aloe bicomitum L.C.Leach
Aloe bulbicaulis Christian
Aloe cameronii Hemsl.
Aloe cameronii var. **cameronii**
Aloe chabaudii Schönland
Aloe chabaudii var. **chabaudii**
Aloe christianii Reynolds
Aloe crassipes Baker
Aloe cryptopoda Baker
Aloe duckeri Christian
Aloe enotata L.C.Leach
Aloe esculenta L.C.Leach
Aloe excelsa A.Berger
Aloe excelsa var. **excelsa**
Aloe fimbrialis S.Carter
Aloe littoralis Baker
Aloe luapulana L.C.Leach
Aloe milne-redheadii Christian
Aloe mzimbana Christian

ZAMBIA (Continued)

Aloe nuttii Baker
Aloe veseyi Reynolds
Aloe zebrina Baker

ZIMBABWE / ZIMBABWE (LE) / ZIMBABWE

Aloe aculeata Pole-Evans
Aloe arborescens Mill.
Aloe ballii Reynolds
Aloe ballii var. ballii
Aloe ballii var. makurupiniensis Ellert
Aloe cameronii Hemsl.
Aloe cameronii var. bondana Reynolds
Aloe cameronii var. cameronii
Aloe carnea S.Carter
Aloe chabaudii Schönland
Aloe chabaudii var. chabaudii
Aloe chabaudii var. verekeri Christian
Aloe christianii Reynolds
Aloe collina S.Carter
Aloe cryptopoda Baker
Aloe excelsa A.Berger
Aloe excelsa var. excelsa
Aloe globuligemma Pole-Evans
Aloe greatheadii Schönland
Aloe greatheadii var. greatheadii
Aloe hazeliana Reynolds
Aloe howmanii Reynolds
Aloe inyangensis Christian
Aloe inyangensis var. inyangensis
Aloe inyangensis var. kimberleyana S.Carter
Aloe littoralis Baker
Aloe munchii Christian
Aloe musapana Reynolds
Aloe myriacantha (Haw.) Schult. & Schult.f.
Aloe ortholopha Christian & Milne-Redh.
Aloe pretoriensis Pole-Evans
Aloe rhodesiana Rendle
Aloe spicata L.f.
Aloe swynnertonii Rendle
Aloe wildii (Reynolds) Reynolds
Aloe zebrina Baker
Pachypodium lealii Welw.
Pachypodium lealii ssp. saundersii (N.E.Br.) G.D.Rowley

PART IV: ACCEPTED TAXA
Further details on homotypic names, basionyms and heterotypic synonyms

The symbol ≡ links homotypic, i.e. the basionym and all later combinations. The abbreviation **incl.** is used for heterotypic synonyms, i.e. synonyms based on interpretation. All synonyms are given in *italic* typeface. Invalid, illegitimate or rejected names are identified by one of the abbreviations *nom. illeg.*, *nom. reject.*, and the article(s) of the *International Code of Botanical Nomenclature* (ICBN) transgressed are indicated in brackets.

QUATRIEME PARTIE: TAXONS ACCEPTES
Autres indications sur les noms homotypiques, les basionymes et les synonymes hétérotypiques

Le symbole "≡" relie le nom homotypique, c'est-à-dire le basionyme, et toutes les combinaisons ultérieures. L'abréviation "**incl.**" est utilisée pour les synonymes hétérotypiques, c'est-à-dire les synonymes fondés sur l'interprétation. Tous les synonymes sont en *italiques*. Les noms non valables, illégitimes ou rejetés sont signalés par une des abréviations "*nom. illeg.*" ou "*nom. reject.*"; les articles du *Code international de nomenclature botanique* (CINB) transgressés figurent entre parenthèses.

PARTE IV: TAXA ACEPTADOS
Mayor información sobre nombres homotípicos, basionímicos y sinónimos heterotípicos

El signo ≡ vincula los homotípicos, es decir, el basionímico y todas las combinaciones más recientes. La abreviatura **incl.** se utiliza para los sinónimos heterotípicos, es decir, los sinónimos basados en interpretaciones. Todos los sinónimos aparecen en *cursiva*. Los nombres inválidos, ilegítimos o rechazados se indican con la abreviatura "*nom. illeg.*" o "*nom. reject.*", y el/los artículo(s) violado(s) del *Código Internacional de Nomenclatura Botánica* (ICBN) aparecen entre paréntesis.

PART IV: ACCEPTED TAXA - ALOE
Further details on homotypic names, basionyms and heterotypic synonyms

QUATRIEME PARTIE: TAXONS ACCEPTES - ALOE
Autres indications sur les noms homotypiques, les basionymes et les synonymes
hétérotypiques

PARTE IV: TAXA ACEPTADOS - ALOE
Mayor información sobre nombres homotípicos, basionímicos y sinónimos heterotípicos

A. aageodonta L.E.Newton 1993, Cact. Succ. J. (US) 65(3): 138-140, ills..
Distribution: Kenya.

A. abyssicola Lavranos & Bilaidi 1971, Cact. Succ. J. (US) 43(5): 204-208, ills..
Distribution: Yemen.

A. aculeata Pole-Evans 1915, Trans. Roy. Soc. South Afr. 5: 34.
Distribution: South Africa (Northern Prov.), Zimbabwe.

A. acutissima H.Perrier 1926, Mém. Soc. Linn. Normandie, Bot. 1(1): 17.
Distribution: Madagascar.

A. acutissima var. **acutissima**
Distribution: Madagascar.

A. acutissima var. **antanimorensis** Reynolds 1956, J. South Afr. Bot. 22(1): 27-29, ills..
Distribution: Madagascar.

A. adigratana Reynolds 1957, J.South Afr. Bot. 23(1): 1-3, ills..
Incl. *Aloe abyssinica* Hook.f. 1900 (*nom. illeg.,* Art. 53:1); incl. *Aloe eru* var. *hookeri*
A.Berger 1908.
Distribution: Ethiopia

A. affinis A.Berger 1908, In Engler, A. (ed.), Pflanzenr. IV.38 (Heft 33): 206.
Distribution: South Africa (Mpumalanga).

A. africana Mill. 1768 Gard. Dict., Ed. 8, [no. 4].
≡ *Aloe perfoliata* var. β L.f. 1753 ≡ *Pachidendron africanum* (Mill.) Haw. 1821; incl. *Aloe
perfoliata* var. *africana* Aiton 1789; incl. *Aloe pseudoafricana* Salm-Dyck 1817; incl. *Aloe
africana* var. *angustior* Haw. 1819; incl. *Aloe africana* var. *latifolia* Haw. 1819; incl. *Aloe
angustifolia* Haw. 1819 ≡ *Pachidendron angustifolium* (Haw.) Haw. 1821; incl. *Pachidendron
africanum* var. *angustum* Haw. 1821; incl. *Pachidendron africanum* var. *latum* Haw. 1821;
incl. *Aloe bolusii* Baker 1880.
Distribution: South Africa (Eastern Cape).

A. ahmarensis Favell, M.B.Mill. & Al-Gifri 1999 CSJA 71(5): 257-259, ills..
Distribution: Yemen

A. albida (Stapf) Reynolds 1947, J. South Afr. Bot. 13(2): 101, 103.
≡ *Leptaloe albida* Stapf 1933; incl.*Aloe kraussii* var. *minor* Baker 1896 ≡ *Aloe myriacantha*
var. *minor* (Baker) A.Berger 1908; incl. *Aloe kraussii* Schönland 1903 (*nom. illeg.,* Art. 53.1).
Distribution: South Africa (Mpumalanga).

A. albiflora Guillaumin 1940, Bull. Mus. Nation. Hist. Nat., Sér. 2, 12: 353.
≡ *Guillauminia albiflora* (Guillaumin) A.Bertrand 1956.
Distribution: Madagascar.

A. albovestita S.Carter & Brandham 1983, Bradleya 1: 20-21, ills..
Distribution: Somalia.

A. aldabrensis (Marais) L.E.Newton & G.D.Rowley 1997, Excelsa 17: 59.
≡ *Lomatophyllum aldabrense* Marias 1975.
Distribution: Seychelles (Aldabra Archipelago, Astove Atoll).

A. alfredii Rauh 1990, Cact. Succ. J. (US) 62(5): 232-233, ills..
Distribution: Madagascar (C).

A. alooides (Bolus) Druten 1956, Bothalia 6(3): 544-545.
≡ *Urginea alooides* Bolus 1881 ≡ *Notosceptrum alooides* (Bolus) Benth. 1883; **incl.** *Aloe recurvifolia* Groenew. 1935.
Distribution: South Africa (Mpumalanga).

A. ambigens Chiov. 1928, Pl. Nov. Min. Not. Ethiop. 1: 6.
Distribution: Somalia.

A. amicorum L.E.Newton 1991, Cact. Succ. J. (US) 63(2): 80-81, ills..
Distribution: Kenya.

A. amudatensis Reynolds 1956, J. South Afr. Bot. 22(3): 136-137, ills..
Distribution: Kenya, Uganda.

A. andongensis Baker 1878, Trans. Linn. Soc. London, Bot.
Distribution: Angola.

A. andongensis var. **andogensis**
Distribution: Angola.

A. andongensis var. **repens** L.C.Leach 1974, J. South Afr. 40(2): 115-116, ills..
Distribution: Angola.

A. andringitrensis H.Perrier 1926, Mém. Soc. Linn. Normandie, Bot. 1(1): 41.
Distribution: Madagascar.

A. angelica Pole-Evans 1934, Flow. Pl. South Afr. 14: t. 554 + text.
Distribution: South Africa (Northern Prov.).

A. angolensis Baker 1878, Trans. Linn. Soc. London, Bot. 1: 236.
Distribution: Angola.

A. anivoranoensis (Rauh & Hebding) L.E.Newton & G.D.Rowley 1998, Bradleya 16: 114.
≡ *Lomatophyllum anivoranoense* Rauh & Hebding 1998.
Distribution: Madagascar (NE).

A. ankaranensis Rauh & Mangelsdorff 2000, in Kakt. And. Sukk. 51(10): 273, Abb. 1-7.
Distribution: Madagascar (NW).

A. ankoberensis M.G.Gilbert & Sebsebe 1997, Kew Bull. 52(1): 146-147.
Distribution: Ethiopia.

A. antandroi (Decary) H.Perrier 1926, Mém. Soc. Linn. Normandie, Bot. 1(1): 19.
≡ *Gasteria antandroi* Decary 1921; **incl.** *Aloe leptocaulon* Bojer 1837 (*nom. nud.*, Art. 32.1c).
Distribution: Madagascar.

A. antsingyensis (Léandri) L.E.Newton & G.D.Rowley 197, Excelsa 17: 59.
≡ *Lomatophyllum antsingyense* Léandri 1935.
Distribution: Madagascar.

A. arborescens Mill. 1768, Gard. Dict., Ed. 8, [no. 3].
Incl. *Aloe perfoliata* var. η L. 1753; incl. *Aloe fruiticosa* Lam. 1783; incl. *Aloe arborea*
Medik. 1783 (nom. illeg., Art. 53.1); incl. *Aloe perfoliata* var. *arborescens* Aiton 1789; incl.
Catevala arborescens Medik. 1789; incl. *Aloe frutescens* Salm-Dyck 1817 ≡ *Aloe arborescens*
var. *frutescens* (Salm-Dyck) Link 1821; incl. *Aloe natalensis* J.M.Wood & M.S.Evans 1901 ≡
Aloe arborescens var. *natalensis* (J.M.Wood & M.S.Evans) A.Berger 1908; incl. *Aloe*
arborescens var. *milleri* A.Berger 1908; incl. *Aloe arborescens* var. *pachythrysa* A.Berger
1908.
Distribution: Malawi, Mozambique, South Africa, Zimbabwe.

A. archeri Lavranos 1977, Cact. Succ. J. (US) 49(2): 74-75.
Distribution: Kenya.

A. arenicola Reynolds 1938, J. South Afr. Bot. 4(1): 21-24, ills..
Distribution: South Africa (Northern Cape, Western Cape).

A. argenticauda Merxm. & Giess 1974, Mitt. Bot. Staatssamml. München 11: 437-444.
Distribution: Namibia

A. aristata Haw. 1825, Philos. Mag. J. 66: 280.
Incl. *Aloe longiaristata* Schult. & Schult.f. 1829; incl. *Aloe aristata* var. *leiophylla* Baker
1880; incl. *Aloe aristata* var. *parvifolia* Baker 1896; incl. *Aloe ellenbergeri* Guillaumin 1934.
Distribution: Lesotho, South Africa (Northern Cape, Western Cape, Eastern Cape, Free State,
KwaZulu-Natal).

A. armatissima Lavranos & Collen. 2000, in Cact. Succ. J. (USA), 72(1): 22
Distribution: Saudi Arabia

A. asperifolia A.Berger 1905, Bot. Jahrb. Syst. 36: 63.
Distribution: Namibia.

A. babatiensis Christian & I.Verd. 1954, Bothalia 6(2): 440-442, ills..
Distribution: Tanzania (United Republic of).

A. bakeri Scott-Elliot 1891, J. Linn. Soc. Bot. 29: 60.
≡ *Guillauminia bakeri* (Scott-Elliot) P.V.Heath 1994.
Distribution: Madagascar.

A. ballii Reynolds 1964, J. South Afr. Bot. 30(3): 123-125, ills..
Distribution: Mozambique, Zimbabwe.

A. ballii var. **ballii**
Distribution: Mozambique, Zimbabwe.

A. ballii var. **makurupiniensis** Ellert 1998, Cact. Succ. J. (US) 70(3): 130-131, ills..
Distribution: Mozambique, Zimbabwe

A. ballyi Reynolds 1953, J. South Afr. Bot. 19(1): 2.
Distribution: Kenya, Tanzania (United Republic of).

A. barberae Dyer 1874, Gard. Chron., ser. Nov. 1: 568.
Incl. *Aloe bainesii* Dyer 1874 ≡ *Aloe bainesii* var. *barberae* (Dyer) Baker 1896.
Distribution: Mozambique, South Africa (KwaZulu-Natal), Swaziland.

A. barbertoniae Pole-Evans 1917, Trans. Roy. Soc. South Afr. 5: 705.
Distribution: South Africa (Mpumalanga).

A. bargalensis Lavranos 1973, Cact. Succ. J. (US) 45(3): 116-117, ills..
Distribution: Somalia.

A. belavenokensis (Rauh & Gerold) L.E.Newton & G.D.Rowley 1997, Excelsa 17: 59.
≡ *Lomatophyllum belavenokense* Rauh & Gerold 1994.
Distribution: Madagascar.

A. bella G.D.Rowley 1974, Repert. Pl. Succ. 23: 12.
≡ *Aloe pulchra* Lavranos 1973 (*nom. illeg.*, Art. 53.1).
Distribution: Somalia.

A. bellatula Reynolds 1956, J. South Afr. Bot. 22(3): 132-134, ills..
≡ *Guillauminia bellatula* (Reynolds) P.V.Heath 1994.
Distribution: Madagascar.

A. berevoana Lavranos 1998, Kakt. and Sukk. 49(7): 161-162, ills..
Distribution: Madagascar (W), sandstone cliffs near sea level.

A. bernadettae J.-B.Castillon 2000, Adansonia, Sér. 3. 22(1): 136, f 1(D-E)
Distribution: Madagascar

Aloe bertemariae Sebsebe & Dioli

Distribution: Ethiopia

A. betsileensis H.Perrier 1926, Mém. Soc. Linn. Normandie, Bot. 1(1): 48.
Distribution: Madagascar.

A. bicomitum L.C.Leach 1977, Kirkia 10: 385-386
≡ *Aloe venusta* Reynolds 1959 (*nom. illeg.*, Art. 53.1).
Distribution: Tanzania (United Republic of), Zambia.

A. boiteaui Guillaumin 1942, Bull. Mus. Nation. Hist. Nat., Sér. 2, 14: 349.
≡ *Lemeea boiteaui* (Guillaumin) P.V.Heath 1994.
Distribution: Madagascar (S) (Toliara: near Fort Dauphin).

A. boscawenii Christian 1942, J. South Afr. Bot. 8(2): 165-167, ills..
Distribution: Tanzania (United Republic of).

A. bosseri J.-B.Castillon 2000, in Adansonia, 22(1): 135
Distribution: Madagascar

A. bowiea Schult. & Schult.f. 1829, Syst. Veg. 7(1): 704.
≡ *Bowiea africana* Haw. 1824 ≡ *Chamaealoe africana* (Haw.) A.Berger 1905.
Distribution: South Africa (Eastern Cape).

A. boylei Baker 1892, Bull. Misc. Inform. [Kew] 1892: 84.
Incl. *Aloe micracantha* Pole-Evans 1923 (*nom. illeg.*, Art. 53.1); **incl.** *Aloe agrophila*
Reynolds 1936.
Distribution: South Africa.

A. boylei ssp. **boylei**
Distribution: South Africa (Eastern Cape, KwaZulu-Natal, Mpumalanga, Northern Prov.).

A. boylei ssp. **major** Hilliard & B.L.Burtt 1985, notes Roy. Bot. Gard. Edinburgh 42(2): 252.
Distribution: South Africa (KwaZulu-Natal).

A. brachystachys Baker 1895, Curtis's Bot. Mag. 121: t. 7399 + text.
Incl. *Aloe lastii* Baker 1901; **incl.** *Aloe schliebenii* Lavranos 1970.
Distribution: Tanzania (United Republic of).

A. branddraaiensis Groenew. 1940, Flow. Pl. South Afr. 20: t. 761 + text.
Distribution: South Africa.

A. brandhamii S. Carter 1994, Fl. Trop. East. Afr., Aloaceae, 32-33, ills..
Distribution: Tanzania (United Republic of)

A. brevifolia Mill. 1771, Gard. Dict. Abr. Ed. 6, [no. 8].
Distribution: South Africa.

A. brevifolia var. **brevifolia**
Incl. *Aloe prolifera* Haw. 1804.
Distribution: South Africa (Western Cape).

A. brevifolia var. **depressa** (Haw.) Baker 1880, J. Linn. Soc., Bot. 18: 160.
≡ *Aloe depressa* Haw. 1804; **incl.** *Aloe perfoliata* var. ζ L. 1753; **incl.** *Aloe serra* DC. 1799 ≡
Aloe brevifolia var. *serra* (DC.) A.Berger 1908.
Distribution: South Africa (Western Cape).

A. brevifolia var. **postgenita** (M.Roem. & Schult.) Baker 1880, J. Linn. Soc., Bot. 18: 160.
≡ *Aloe postgenita* M.Roem. & Schult. 1830; **incl.** *Aloe prolifera* var. *major* Salm-Dyck 1817.
Distribution: South Africa (Western Cape).

A. breviscapa Reynolds & P.R.O.Bally 1958, J. South Afr. Bot. 24(4): 176-177, t. 19.
Distribution: Somalia.

A. broomii Schönland 1907, Rec. Albany Mus. 2: 137.
Distribution: Lesotho, South Africa

A. broomii var. **broomii**
Distribution: Lesotho, South Africa (Eastern Cape, Western Cape, Northern Cape, Free State)

A. broomii var. **tarkaensis** Reynolds 1936, J. South Afr. Bot. 2(2): 72-73, ills..
Distribution: South Africa (Eastern Cape).

A. brunneodentata Lavranos & Collen. 2000, in Cact. Succ. J. (USA), 72(2): 86
Distribution: Saudi Arabia

A. brunneostriata Lavranos & S.Carter 1992, Cact. Succ. J. (US) 64(4): 206-208, ills..
Distribution: Somalia.

A. buchananii Baker 1895, Bull. Misc. Inform. [Kew] 1895: 119.
Distribution: Malawi.

A. buchlohii Rauh 1966, Kakt. And. Sukk. 17(1): 2-4, ills..
Distribution: Madagascar.

A. buettneri A.Berger 1905, Bot. Jahrb. Syst. 36: 60.
Incl. *Aloe barteri* Baker 1880; **incl.** *Aloe barteri* var. *dahomensis* A.Chev. 1952 (*nom. nud.*,
Art. 36.1); **incl.** *Aloe barteri* var. *sudanica* A.Chev. 1952 (*nom. nud.*, Art. 36.1); **incl.** *Aloe
paludicola* A.Chev. 1952 (*nom. nud.*, Art. 36.1).

Distribution: Benin, Ghana, Mali, Nigeria, Togo.

A. buhrii Lavranos 1971, J. South Afr. Bot. 37(1): 37-40, ills..
Distribution: South Africa (Northern Cape).

A. bukobana Reynolds 1955, J. South Afr. Bot. 20(4): 169-171, ills..
Distribution: Tanzania (United Republic of).

A. bulbicaulis Christian 1936, Flow. Pl. South Afr. 16: t. 630 + text.
Incl. *Aloe trothae* A.Berger 1905.
Distribution: Angola, Democratic Republic of the Congo (the), Malawi, Tanzania (United Republic of), Zambia.

A. bulbillifera H.Perrier 1926, Mém. Soc. Linn. Normandie, Bot. 1(1): 22.
Distribution: Madagascar.

A. bulbillifera var. **bulbillifera**
Distribution: Madagascar.

A. bulbillifera var. **paulianae** Reynolds 1956, J. South Afr. Bot. 22(1): 26-27, ills..
Distribution: Madagascar.

A. bullockii Reynolds 1961, J. South Afr. Bot. 27(2): 73-75, ills..
Distribution: Tanzania (United Republic of).

A. burgersfortensis Reynolds 1936, J. South Afr. Bot. 2(1): 31-34, ills..
Distribution: South Africa (Mpumalanga).

A. bussei A.Berger 1908, In Engler, A. (ed.), Pflanzenr. IV.38 (Heft 33): 273.
Incl. *Aloe morogoroensis* Christian 1940.
Distribution: Tanzania (United Republic of).

A. calcairophila Reynolds 1961, J. South Afr. Bot. 27(1): 5-6, ills..
≡ *Guillauminia calcairophila* (Reynolds) P.V.Heath 1994; **incl.** *Aloe calcairophylla* hort. (*nom. nud.*, Art. 61.1).
Distribution: Madagascar.

A. calidophila Reynolds 1954, J. South Afr. Bot. 20(1): 26-28, ills..
Distribution: Ethiopia, Kenya.

A. cameronii Hemsl. 1903, Curtis's Bot. Mag. 1903: t. 7915 + text.
Distribution: Malawi, Mozambique, Zambia, Zimbabwe.

A. cameronii var. **bondana** Reynolds 1966, Aloes Trop. Afr. & Madag., 353, ills..
Distribution: Zimbabwe.

A. cameronii var. **cameronii**
Distribution: Malawi, Mozambique, Zambia, Zimbabwe.

A. cameronii var. **dedzana** Reynolds 1965, J. South Afr. Bot. 31(2): 167-168, ills..
Distribution: Malawi, Mozambique.

A. camperi Schweinf. 1894, Bull. Herb. Boissier 2 (app. 2): 67.
Incl. *Aloe abyssinica* Salm-Dyck 1817 (*nom. illeg.*, Art. 53.1); **incl.** *Aloe spicata* Baker 1896 (*nom. illeg.*, Art. 53.1); **incl.** *Aloe albopicta* hort. *ex* A.Berger 1908; **incl.** *Aloe eru* A.Berger 1908; **incl.** *Aloe eru* var. *cornuta* A.Berger 1908; **incl.** *Aloe eru* forma *erecta* hort. *ex* A.Berger

1908; **incl.** *Aloe eru* forma *glauca* hort. *ex* A.Berger 1908; **incl.** *Aloe eru* forma *maculata* hort. *ex* A.Berger 1908; **incl.** *Aloe eru* forma *parvipunctata* hort. *ex* A.Berger 1908.
Distribution: Eritrea, Ethiopia.

A. canarina S.Carter 1994, Fl. Trop. East Afr., Aloaceae, 41-42, ills..
Distribution: Sudan (the), Uganda.

A. cannellii L.C.Leach 1971, J. South Afr. Bot. 37(1): 41-46, ills..
Distribution: Mozambique

A. capitata Baker 1883, J. Linn. Soc., Bot. 20: 272.
Distribution: Madagascar

A. capitata var. **capitata**
Incl. *Aloe cernua* Tod. 1890
Distribution: Madagascar

A. capitata var. **cipolinicola** H.Perrier 1926, Mém. Soc. Linn. Normandie, Bot. 1(1): 39.
Distribution: Madagascar

A. capitata var. **gneissicola** H.Perrier 1926, Mém. Soc. Linn. Normandie, Bot. 1(1): 37.
Distribution: Madagascar.

A. capitata var. **quartziticola** H.Perrier 1926, Mém. Soc. Linn. Normandie, Bot. 1(1): 38, ills..
Distribution: Madagascar.

A. capitata var. **silvicola** H.Perrier 1926, Mém. Soc. Linn. Normandie, Bot. 1(1): 39.
Distribution: Madagascar.

A. capmanambatoensis Rauh & Gerold 2000, in Kakt. And. Sukk. 51(11): 293-294, pl.
Distribution: Madagascar (NE)

A. carnea S.Carter 1996, Kew Bull. 51(4): 784-785.
Distribution: Zimbabwe

A. castanea Schönland 1907, Rec. Albany Mus. 2: 138.
Distribution: South Africa (Gauteng, Mpumalanga, Northern Prov.).

A. castellorum J.R.I.Wood 1983, Kew Bull. 38(1): 25-26, t. 2.
Distribution: Saudi Arabia, Yemen.

A. catengiana Reynolds 1961, Kirkia 1: 160.
Distribution: Angola.

A. cephalophora Lavranos & Collen. 2000, in Cact. Succ. J. (USA), 72(1): 20
Distribution: Saudi Arabia

A. chabaudii Schönland 1905, Gard. Chron., ser. 3, 38: 102.
Distribution: Malawi, Mozambique, Swaziland, South Africa, Tanzania (United Republic of), Democratic Republic of the Congo (the), Zambia, Zimbabwe.

A. chabaudii var. **chabaudii**
Distribution: Malawi, Mozambique, Swaziland, South Africa, Tanzania (United Republic of), Democratic Republic of the Congo (the), Zambia, Zimbabwe.

A. chabaudii var. **mlanjeana** Christian 1938, Flow. Pl. South Afr. 18: t. 698 + text.
Distribution: Malawi.

A. chabaudii var. **verekeri** Christian 1938, Flow. Pl. South Afr. 18: t. 699 + text.
Distribution: Mozambique, Zimbabwe.

A. cheranganiensis S.Carter & Brandham 1979, Cact. Succ. J. Gr. Brit. 41(1): 4-6, ills..
Distribution: Kenya, Uganda.

A. chlorantha Lavranos 1973, J. South Afr. Bot. 39(1): 85-90, ills..
Distribution: South Africa (Northern Cape).

A. chortolirioides A.Berger 1908, In Engler, A. (ed.), Pflanzenr. IV.38 (Heft 33): 171.
Distribution: South Africa, Swaziland.

A. chortolirioides var. **chortolirioides**
Incl. *Aloe boastii* Letty 1934 ≡ *Aloe chortolirioides* var. *boastii* (Letty) Reynolds 1950.
Distribution: South Africa (Mpumalanga), Swaziland.

A. chortolirioides var. **woolliana** (Pole-Evans) Glen & D.S.Hardy 1987, South Afr. J. Bot.
53(6): 489-490.
≡ *Aloe woolliana* Pole-Evans 1934.
Distribution: South Africa (Mpumalanga, Northern Prov.), Swaziland.

A. christianii Reynolds 1936, J. South Afr. Bot. 2(4): 171-173, ills..
Distribution: Angola, Malawi, Mozambique, Tanzania (United Republic of), Democratic
Republic of the Congo (the), Zambia, Zimbabwe.

A. chrysostachys Lavranos & L.E.Newton 1976, Cact. Succ. J. (US) 48 (6): 278-279, ills..
Incl. *Aloe meruana* Lavranos 1980.
Distribution: Kenya.

A. ciliaris Haw. 1825, Philos. Mag. J. 66: 281.
Distribution: South Africa.

A. ciliaris var. **ciliaris**
Incl. *Aloe ciliaris* var. *flanaganii* Schönland 1903 ≡ *Aloe ciliaris* forma *flanaganii*
(Schönland) Resende 1943; **incl.** *Aloe ciliaris* forma *gigas* Resende 1938.
Distribution: South Africa (Eastern Cape).

A. ciliaris var. **redacta** S.Carter 1990, Kew Bull. 45(4): 643.
Distribution: South Africa (Eastern Cape).

A. ciliaris var **tidmarshii** Schönland 1903, Rec. Albany Mus. 1: 41.
≡ *Aloe ciliaris* forma *tidmarshii* (Schönland) Resende 1943 ≡ *Aloe tidmarshii* (Schönland)
F.S.Mull. *ex* R.A.Dyer 1943.
Distribution: South Africa (Eastern Cape).

A. citrea (Guillaumin) L.E.Newton & G.D.Rowley 1997, Excelsa 17: 61.
≡ *Lomatophyllum citreum* Guillaumin 1944.
Distribution: Madagascar.

A. citrina S.Carter & Brandham 1983, Bradleya 1: 21-23, ills..
Distribution: Ethiopia, Kenya, Somalia.

A. classenii Reynolds 1965, J. South Afr. Bot. 31(4): 271-173, ills..
Distribution: Kenya.

A. claviflora Burch. 1822, Trav. South. Afr. 1: 272.
Incl. *Aloe schlechteri* Schönland 1903; **incl.** *Aloe decora* Schönland 1905.

Distribution: Namibia, South Africa (Northern Cape, Western Cape, Eastern Cape, Free State).

A. colletteae Lavranos 1995, Cact. Succ. J. (US) 67(1): 32-33, ills..
Distribution: Oman (Dhofar Prov.).

A. collina S.Carter 1996, Kew Bull. 51(4): 781-782.
Distribution: Zimbabwe.

A. commixta A.Berger 1908, in Engler, A. (ed.) Pflanzenr. IV.38 (Heft 33): 260-261, ills..
Incl. *Aloe perfoliata* var. α L. 1753; **incl.** *Aloe gracilis* Baker 1880 (*nom. illeg.,* Art. 53.1).
Distribution: South Africa (Western Cape).

A. comosa Marloth & A.Berger 1905, Bot. Jahrb. Syst. 38: 86.
Distribution: South Africa (Western Cape).

A. compressa H.Perrier 1926, Mém. Soc. Linn. Normandie, Bot. 1(1): 33.
Distribution: Madagascar.

A. compressa var. **compressa**
Distribution: Madagascar.

A. compressa var. **paucituberculata** Lavranos 1998, Kakt. And. Sukk. 49(7): 158-159, ills..
Distribution: Madagascar (C).

A. compressa var. **rugosquamosa** H.Perrier 1926, Mém. Soc. Linn. Normandie, Bot. 1(1): 34.
Distribution: Madagascar.

A. compressa var. **schistophila** H.Perrier 1926, Mém. Soc. Linn. Normandie, Bot. 1(1): 34.
Distribution: Madgascar.

A. comptonii Reynolds 1950, Aloes South Afr. [ed. 1], 382-385, ills..
Distribution: South Africa (Western Cape, Eastern Cape).

A. confusa Engl. 1895, Pfl. Welt Ost-Afr., Teil C, 141.
Distribution: Kenya, Tanzania (United Republic of).

A. congdonii S.Carter 1994, Fl. Trop. East Afr., Aloaceae, 18, 20, ill. (p.10).
Distribution: Tanzania (United Republic of).

A. conifera H.Perrier 1926, Mém. Soc. Linn. Normandie, Bot. 1(1): 47.
Distribution: Madagascar.

A. cooperi Baker 1874, Gard. Chron., ser. Nov. 1: 628.
Distribution: South Africa, Swaziland.

A. cooperi ssp. **cooperi**
Incl. *Aloe schmidtiana* Regel 1879.
Distribution: South Africa (KwaZulu-Natal, Mpumalanga), Swaziland.

A. cooperi ssp. **pulchra** Glen & D.S.Hardy 1987, Flow. Pl. Afr. 49(3-4): t. 1944 + text.
Distribution: South Africa (KwaZulu-Natal), Swaziland.

A. corallina I.Verd. 1979, Flow. Pl. Afr. 45: t. 1788 + text.
Distribution: Namibia.

A. crassipes Baker 1880, J. Linn. Soc., Bot. 18: 162.
Distribution: Sudan (the), Zambia.

A. cremersii Lavranos 1974, Adansonia, n.s., 14(1): 99-101, ills..
Distribution: Madagascar.

A. cremnophila Reynolds & P.R.O.Bally 1961, J. South Afr. Bot. 27(2): 77-79, t. 13-14.
Distribution: Somalia.

A. cryptoflora Reynolds 1965, J. South Afr. Bot. 31(4): 281-284, ills..
Distribution: Madagascar.

A. cryptopoda Baker 1884, J. Bot. 1884: 52.
Incl. *Aloe wickensii* Pole-Evans 1915; **incl.** *Aloe wickensii* var. *wickensii*; **incl.** *Aloe pienaarii* Pole-Evans 1915; **incl.** *Aloe wickensii* var. *lutea* Reynolds 1935.
Distribution: Botswana, Malawi, Mozambique, South Africa (Northern Prov., North-West Prov., Mpumalanga), Swaziland, Zambia, Zimbabwe.

A. cyrtopylla Lavranos 1998, Kakt. And Sukk. 49(7): 159-161, ills..
Distribution: Madagascar (S).

A. dabenorisana Van Jaarsv. 1982, J. South Afr. Bot. 48(3): 419-424, ills..
Distribution: South Africa (Northern Cape).

A. dawei A.Berger 1906, Notizbl. Königl. Bot. Gart. Berlin 4: 246.
Incl. *Aloe beniensis* De Wild. 1921; **incl.** *Aloe pole-evansii* Christian 1940.
Distribution: Democratic Republic of the Congo (the), Kenya, Rwanda, Uganda.

A. debrana Christian 1947, Flow. Pl. Afr. 26: t. 1016 + text.
Incl. *Aloe berhana* Reynolds 1957.
Distribution: Ethiopia.

A. decorsei H.Perrier 1926, Mém. Soc. Linn. Normandie, Bot. 1(1): 43.
Distribution: Madagascar.

A. decurva Reynolds 1957, J. South Afr. Bot. 23(1): 15-17, ills..
Distribution: Mozambique.

A. delphinensis Rauh 1990, Cact. Succ. J. (US) 62(5): 230-232, ills..
Distribution: Madagascar.

A. deltoideodonta Baker 1883, J. Linn. Soc., Bot. 20: 271.
Distribution: Madagascar.

A. deltoideodonta var. **brevifolia** H. Perrier 1926, Mém. Soc. Linn. Normandie, Bot. 1(1): 24.
Distribution: Madagascar.

A. deltoideodonta var. **candicans** H.Perrier 1926, Mém. Soc. Linn. Normandie, Bot. 1(1): 25.
Distribution: Madagascar.

A. deltoideodonta var. **deltoideodonta**
Incl. *Aloe deltoideodonta* var. *typica* H.Perrier 1926 (*nom. nud.*, Art. 24.3).
Distribution: Madagascar.

A. descoingsii Reynolds 1958, J. South Afr. Bot. 24(2): 103-105 ills..
≡ *Guillauminia descoingsii* (Reynolds) P.V.Heath 1994.
Distribution: Madagascar.

A. descoingsii ssp. **augustina** Lavranos 1995, Cact. Succ. J. (US) 67(3): 158-161, ills..
Distribution: Madagascar.

A. descoingsii ssp. **descoingsii**
Distribution: Madagascar.

A. deserti A.Berger 1905, Bot. Jahrb. Syst. 36: 61.
Distribution: Kenya, Tanzania (United Republic of).

A. dewetii Reynolds 1937, J. South Afr. Bot. 3(3): 139-141, ills..
Distribution: South Africa (KwaZulu-Natal, Mpumalanga), Swaziland.

A. dewinteri Giess 1973, Bothalia 11: 120-122 ills..
Distribution: Namibia.

A. dhufarensis Lavranos 1967, Cact. Succ. J. (US) 39(5): 167-171, ills..
Distribution: Oman.

A. dichotoma Masson 1776, Philos. Trans. 66: 310.
≡ *Rhipidondenrum dichotomum* (Masson) Willd. 1811; **incl.** *Aloe ramosa* Haw. 1804; **incl.**
Aloe montana Schinz 1896 ≡ *Aloe dichotoma* var. *montana* (Schinz) A.Berger 1908.
Distribution: Namibia, South Africa (Northern Cape).

A. dinteri A.Berger 1914, in Dinter, Neue Pfl. Deutsch-SWA, 14.
Distribution: Namibia.

A. diolii L.E.Newton 1995, Cact. Succ. J. (US) 67(5): 277-279, ills., SEM-ills..
Distribution: Sudan (the).

A. distans Haw. 1812, Synops. Pl. Succ., 78.
Incl. *Aloe mitriformis* var. *angustior* Lam. 1784; **incl.** *Aloe perfoliata* var. *brevifolia* Aiton
1789; **incl.** *Aloe brevifolia* Haw. 1804 (*nom illeg.*, Art. 53.1); **incl.** *Aloe mitriformis* var.
brevifolia Aiton 1810.
Distribution: South Africa (Western Cape).

A. divaricarta A.Berger 1905, Bot. Jahrb. Syst. 36: 65.
Distribution: Madagascar.

A. divaricata var. **divaricata**
Incl. *Aloe sahundra* Bojer 1837 (*nom. nud.*, Art. 32.1c); **incl.** *Aloe vahontsohy* Decorse *ex*
Poiss. 1912; **incl.** *Aloe vaotsohy* Decorse & Poiss. 1912; **incl.** *Aloe vahontsohy* H.Perrier 1938
(*nom. nud.*).
Distribution: Madagascar.

A. divaricata var. **rosea** (Decary) Reynolds 1958, Aloes Madag. Revis., 133.
≡ *Aloe vaotsohy* var. *rosea* Decary 1921.
Distribution: Madagascar.

A. doei Lavranos 1965, J. South Afr. Bot. 31(2): 163-166, ills..
Distribution: Yemen.

A. doei var. **doei**
Distribution: Yemen (Audhali escarpment).

A. doei var. **lavranosii** Marn.-Lap. 1970, Cact. Succ. J. (US) 42(6): 262, ills..
Distribution: Yemen.

A. dominella Reynolds 1938, J. South Afr. Bot. 4(4): 101-103, ills..
Distributions: South Africa (KwaZulu-Natal).

A. dorotheae A.Berger 1908, in Engler, A. (ed.), Pflanzenr. IV.38 (Heft 33): 263-264.
Incl. *Aloe harmsii* A.Berger 1908.
Distribution: Tanzania (United Republic of).

A. duckeri Christian 1940, J. South Afr. Bot. 6(4): 179-180, ills..
Distribution: Malawi, Tanzania (United Republic of), Zambia.

A. dyeri Schönland 1905, Rec. Albany Mus. 1: 289.
Distribution: South Africa (Mpumalanga).

A. ecklonis Salm-Dyck 1849, Monogr. Gen. Aloes & Mesembr. 5: t. 5 + text.
Distribution: Lesotho, South Africa (Eastern Cape, KwaZulu-Natal, Free State, Mpumalanga), Swaziland.

A. edentata Lavranos & Collen.
Distribution: Saudi Arabia

A. elata S.Carter & L.E.Newton 1994, in S.Carter, Fl. Trop. East Afr. Aloaceae, 56-57, t. 4: lower right.
Distribution: Kenya, Tanzania (United Republic of).

A. elegans Tod. 1882, Hort. Bot. Panorm. 2: 25, t 29.
Incl. *Aloe abyssinica* var. *peacockii* Baker 1880; **incl.** *Aloe vera* var. *aethiopica* Schweinf. 1894 ≡ *Aloe aethiopica* (Schweinf.) A.Berger 1905; **incl.** *Aloe schweinfurthii* Hook.f. 1899 (*nom. illeg.*, Art. 53.1); **incl.** *Aloe peacockii* A.Berger 1905; **incl.** *Aloe percrassa* var. *saganeitiana* A. Berger 1908; **incl.***Aloe abyssinica* A.Berger 1908 (*nom. illeg.*, Art. 53.1).
Distribution: Eritrea, Ethiopia.

A. elgonica Bullock 1932, Bull. Misc. Inform. [Kew] 1932: 503.
Distribution: Kenya.

A. ellenbeckii A.Berger 1905, Bot. Jahrb. Syst. 36: 59.
Incl. *Aloe dumetorum* B.Mathew & Brandham 1977.
Distribution: Ethiopia, Kenya, Somalia.

A. eminens Reynolds & P.R.O.Bally 1958, J. South Afr. Bot. 24: 187-189, t. 30-32.
Distribution: Somalia.

A. enotata L.C.Leach 1972, J. South Afr. Bot. 38(3): 187-193, ills..
Distribution: Zambia.

A. eremophila Lavranos 1965, J. South Afr. Bot. 31(1): 71-74, ills..
Distribution: Yemen (Hadhramaut).

A. erensii Christian 1940, Flow. Pl. South Afr. 20: t. 797 + text.
Distribution: Kenya, Sudan (the).

A. ericetorum Bosser 1968, Adansonia, n.s., 8(4): 508-509, ills..
Distribution: Madagascar (C).

A. erythrophylla Bosser 1968, Adansonia, n.s., 8(4): 508-511, ill..
Distribution: Madagascar (C).

A. esculenta L.C.Leach 1971, J. South Afr. Bot. 37(4): 249-259, ills..
  Distribution: Angola, Botswana, Namibia, Zambia.

A. eumassawana S.Carter & al. 1996, Kew Bull. 51(4): 776.
  Distribution: Eritrea.

A. excelsa A.Berger 1906, Notizbl. Königl. Bot. Gart. Berlin 4: 247.
  Distribution: Mozambique, Malawi, South Africa, Zambia, Zimbabwe.

A. excelsa var. **breviflora** L.C.Leach 1977, Kirkia 10 (2): 387-389, ill..
  Distribution: Malawi, Mozambique.

A. excelsa var. **excelsa**
  Distribution: Malawi, Mozambique, South Africa (Northern Prov.), Zambia, Zimbabwe.

A. falcata Baker 1880, J. Linn. Soc., Bot. 18: 181.
  Distribution: South Africa (Northern Cape, Western Cape).

A. ferox Mill. 1768, Gard. Dict., Ed. , no. 22.
  Incl. *Aloe perfoliata* var. ε L. 1753; **incl.** *Aloe perfoliata* var. γ L. 1753 ≡ *Pachidendron ferox*
  (Mill.) Haw. 1821; **incl.** *Aloe socotorina* Masson 1773; **incl.** *Aloe perfoliata* Thunb. 1785;
  incl. *Aloe perfoliata* var. *ferox* Aiton 1789; **incl.** *Aloe perfoliata* var. ζ Willd. 1799; **incl.** *Aloe*
  muricata Haw. 1804; **incl.** *Aloe supralaevis* Haw. 1804 ≡ *Pachidendron supralaeve* (Haw.)
  Haw. 1821; **incl.** *Aloe pseudoferox* Salm-Dyck 1817 ≡ *Pachidendron pseudoferox* (Salm-
  Dyck) Haw. 1821; **incl.** *Aloe subferox* Spreng. 1826 ≡ *Aloe ferox* var. *subferox* (Spreng.)
  Baker 1880; **incl.** *Aloe ferox* var. *incurva* Baker 1880; **incl.** *Aloe ferox* var. *hanburyi* Baker
  1896; **incl.** *Aloe galpinii* Baker 1901 ≡ *Aloe ferox* var. *galpinii* (Baker) Reynolds 1937; **incl.**
  Aloe candelabrum A.Berger 1906; **incl.** *Aloe ferox* var. *erythrocarpa* A.Berger 1908.
  Distribution: Lesotho, South Africa (Western Cape, Eastern Cape, Free State, KwaZulu-
  Natal).

A. fibrosa Lavranos & L.E.Newton 1976, Cact. Succ. J. (US) 48(6): 273-275, ills..
  Distribution: Kenya, Tanzania (United Republic of).

A. fievetii Reynolds 1965, J. South Afr. Bot. 31(4): 279-281, ills..
  Distribution: Madagascar.

A. fimbrialis S.Carter 1996, Kew Bull. 51(4): 779-781, ills..
  Distribution: Tanzania (United Republic of), Zambia.

A. fleuretteana Rauh & Gerold 2000, in Kakt. And. Sukk., 51(5): 122
  Distribution: Madagascar

A. fleurentiniorum Lavranos & L.E.Newton 1966, Cact. Succ. J. (US) 49(3): 113-114, ills..
  Distribution: Yemen.

A. flexilifolia Christian 1942, J. South Afr. Bot. 8(2): 167-169, ills..
  Distribution: Tanzania (United Republic of).

A. forbesii Balf.f. 1903, in Forbes & Ogilvie-Grant, Nat. Hist. Socotra, 511-512, t. 26B.
  Distribution: Yemen (Socotra).

A. fosteri Pillans 1933, South Afr. Gard. 23: 140.
  Distribution: South Africa (Mpumalanga).

A. fouriei D.S.Hardy & Glen 1987, Flow. Pl. Afr. 49(3-4): t. 1941 + text.
  Distribution: South Africa (Mpumalanga).

A. fragilis Lavranos & Röösli 1994, Cact. Succ. J. (US) 66(1): 45 ills..
Distribution: Madagascar (N).

A. framesii L.Bolus 1933, South Afr. Gard. 1933: 140 (June).
Incl. *Aloe amoena* Pillans 1933.
Distribution: South Africa (Northern Cape, Western Cape).

A. francombei L.E.Newton 1994, Brit. Cact. Succ. J. 12(2): 54-55, ills..
Distribution: Kenya.

A. friisii Sebese & M.G.Gilbert
Distribution: Ethiopia

A. fulleri Lavranos 1967, Cact. Succ. J. (US) 39(4): 125-127, ills..
Distribution: Yemen.

A. gariepensis Pillans 1933, South Afr. Gard. 23: 213.
Incl. *Aloe gariusiana* Dinter 1928 (*nom. nud.*, Art. 32.1c).
Distribution: Namibia, South Africa (N) (Northern Cape).

A. gerstneri Reynolds 1937, J. South Afr. Bot. 3: 133.
Distribution: South Africa (Kwa-Zulu-Natal).

A. gilbertii T.Reynolds *ex* Sebsebe & Brandham 1992, Kew Bull. 47(3): 509, 512, ills..
Distribution: Ethiopia.

A. gilbertii spp. **gilbertii**
Distribution: Ethiopia.

A. gilbertii ssp. **megalacanthoides** M.G.Gilbert & Sebsebe 1997, Kew Bull. 52(1): 151.
Distribution: Ethiopia.

A. gillettii S.Carter 1994, Kew. Bull. 49(3): 417.
Distribution: Somalia.

A. glabrescens (Reynolds & P.R.O.Bally) S.Carter & Brandham 1983, Bradleya 1: 23-24, ills..
≡ *Aloe rigens* var. *glabrescens* Reynolds & P.R.O.Bally 1958.
Distribution: Somalia.

A. glauca Mill. 1768, Gard. Dict., Ed. 8, no. 16.
Distribution: South Africa.

A. glauca var. **glauca**
Incl. *Aloe perfoliata* var. κ L. 1753; **incl.** *Aloe perfoliata* var. *glauca* Aiton 1789; **incl.** *Aloe rhodacantha* DC. 1799; **incl.** *Aloe glauca* var. *major* Haw. 1812; **incl.** *Aloe glauca* var. *minor* Haw. 1812; **incl.** *Aloe glauca* var. *elatior* Salm-Dyck 1817; **incl.** *Aloe glauca* var. *humilior* Salm-Dyck 1817.
Distribution: South Africa (Northern Cape, Western Cape).

A. glauca var. **spinosior** Haw. 1821, Revis. Pl. Succ. 40.
Incl. *Aloe muricata* Schult. 1809 ≡ *Aloe glauca* var. *muricata* (Schult.) Baker 1880.
Distribution: South Africa (Northern Cape, Western Cape).

A. globuligemma Pole-Evans 1915, Trans. Roy. Soc. South Afr. 5: 30.
Distribution: Botswana, South Africa (Northern Prov., Mpumalanga), Zimbabwe.

A. gossweileri Reynolds 1962, J. South Afr. Bot. 28(3): 205-207, ills..
Distribution: Angola.

A. gracilicaulis Reynolds & P.R.O.Bally 1958, J. South Afr. Bot. 24(4): 184-186, t. 27-28.
Distribution: Somalia.

A. gracilis Haw. 1825, Philos. Mag. J. 66: 279.
Distribution: South Africa.

A. gracilis var. **decumbens** Reynolds 1950, Aloes South Afr. [ed. 1], 358-359.
Distribution: South Africa (Western Cape).

A. gracilis var. **gracilis**
Incl. *Aloe laxiflora* N.E.Br. 1906.
Distribution: South Africa (Eastern Cape).

A. grandidentata Salm-Dyck 1822, Observ. Bot. Hort. Dyck. 3.
Distribution: Botswana, South Africa (Northern Cape, North-West Prov., Free State).

A. grata Reynolds 1960, J. South Afr. Bot. 26(2): 87-89, pl. 8-9.
Distribution: Angola.

A. greatheadii Schönland 1904, Rec. Albany Mus. 1: 121.
Distribution: Botswana, Democratic Republic of the Congo (the), Malawi, Mozambique, South Africa, Swaziland, Zimbabwe.

A. greatheadii var. **davyana** (Schönland) Glen & D.S.Hardy 1987, South Afr. J. Bot. 53(6): 490-491.
≡ *Aloe davyana* Schönland 1904; **incl.** *Aloe longibracteata* Pole-Evans 1915; **incl.** *Aloe comosibracteata* Reynolds 1936; **incl.** *Aloe graciliflora* Groenew. 1936; **incl.** *Aloe labiaflava* Groenew. 1936; **incl.** *Aloe mutans* Reynolds 1936; **incl.** *Aloe verdoorniae* Reynolds 1936; **incl.** *Aloe davyana* var. *subolifera* Groenew. 1939.
Distribution: South Africa (Free State, Gauteng, KwaZulu-Natal, Northern Prov., North-West Prov.) Swaziland.

A. greatheadii var. **greatheadii**
Incl. *Aloe pallidiflora* A.Berger 1905; **incl.** *Aloe termetophila* De Wild. 1921.
Distribution: Botswana, Democratic Republic of the Congo (the), Malawi, Mozambique, South Africa (Northern Prov.), Zimbabwe.

A. greenii Baker 1880, J. Linn. Soc. Bot. 18: 165.
Distribution: South Africa (KwaZulu-Natal).

A. grisea S.Carter & Brandham 1983, Bradleya 1: 19-20, ills..
Distribution: Somalia.

A. guerrae Reynolds 1960, J. South AFr. Bot. 26(2): 85-87, pl. 6-7.
Distribution: Angola.

A. guillaumetii Cremers 1976, Adansonia, n.s., 15(4): 498-501, ills..
Distribution: Madagascar.

A. haemanthifolia A.Berger & Marloth 1905, Bot. Jahrb. Syst. 38: 85.
Distribution: South Africa (Western Cape).

A. hardyi Glen 1987, Flow. Pl. Afr. 49(3-4): pl. 1942 + 3 pp. of text.
Distribution: South Africa (Mpumalanga).

A. harlana Reynolds 1957, J. South Afr. Bot. 23(1): t. 9.
Distribution: Ethiopia.

A. haworthioides Baker 1886, J. Linn. Soc., Bot. 22: 259.
≡ *Aloinella haworthioides* (Baker) Lemée 1939 (*nom. nud.*, Art. 43.1) ≡ *Lemeea haworthioides* (Baker) P.V.Heath 1993.
Distribution: Madagascar.

A. haworthioides var. **aurantiaca** H.Perrier 1926, Mém. Soc. Linn. Normandie, Bot. 1(1): 50.
Distribution: Madagascar.

A. haworthioides var. **haworthioides**
Distribution: Madagascar.

A. hazeliana Reynolds 1959, J. South Afr. Bot. 25(4): 279-281, pl. 25-26.
Distribution: Mozambique, Zimbabwe.

A. helenae Danguy 1929, Bull. Mus. Nation. Hist. Nat., Sér. 2, 1: 433.
Distribution: Madagascar (S).

A. heliderana Lavranos 1973, Cact. Succ. J. (US) 45(3): 114-115, ills..
Distribution: Somalia.

A. hemmingii Reynolds & P.R.O.Bally 1964, J. South Afr. Bot. 30(4): 221-222, ills..
Distribution: Somalia.

A. hendrickxii Reynolds 1955, J. South Afr. Bot. 21(2): 51-53, ills..
Distribution: Democratic Republic of the Congo (the).

A. hereroensis Engl. 1888, Bot. Jahrb., Syst. 10: 2.
Incl. *Aloe heroroensis* var. *hereroensis*.
Distribution: Angola, Namibia, South Africa (Northern Cape).

A. heybensis Lavranos 1999, Cact. Succ. J. (US) 71(3): 159-160, ills..
Distribution: Somalia (S).

A. hijazensis Lavranos & Collen. 2000, in Cact. Succ. J. (USA), 72(1): 23
Distribution: Saudi Arabia

A. hildebrandtii Baker 1888, Curtis's Bot. Mag. 1888: t. 6981 + text.
Incl. *Aloe gloveri* Reynolds & P.R.O.Bally 1958.
Distribution: Somalia.

A. hlangapies Groenew. 1936, Tydskr. Wetensk. Kuns 14: 60-63.
Incl. *Aloe hlangapitis* Groenew. 1936 (*nom. nud.*, Art. 61.1); incl. *Aloe hlangapensis* Groenew. 1937 (*nom. nud.*, Art. 61.1).
Distribution: South Africa (KwaZulu-Natal, Mpumalanga).

A. howmanii Reynolds 1961, Kirkia 1: 156-157, t. 15.
Distribution: Zimbabwe.

A. humbertii H.Perrier 1931, Bull. Mus. Nation. Hist. Nat., Sér. 2, 3: 692.
Distribution: Madagascar (S).

A. humilis (L.) Mill. 1771, Gard. Dict. Abr. Ed. 6, no. 10.
≡ *Aloe perfoliata* var. *humilis* L. 1753 ≡ *Catevala humilis* (L.) Medik. 1786; incl. *Aloe humilis* var. *humilis*; incl. *Aloe humilis* var. *incurva* Haw. 1804 ≡ *Aloe incurva* (Haw.) Haw. 1812; incl. *Aloe suberecta* Haw. 1804 ≡ *Aloe humilis* var. *suberecta* (Haw.) Baker 1896; incl. *Aloe*

suberecta var. *acuminata* Haw. 1804 ≡ *Aloe acuminata* (Haw.) Haw. 1812 ≡ *Aloe humilis* var. *acuminata* (Haw.) Baker 1880; **incl.***Aloe tuberculata* Haw. 1804; **incl.** *Aloe humilis* Ker Gawl. 1804 (*nom. illeg.*, Art. 53.1);**incl.** *Aloe echinata* Wildenow 1809 ≡ *Aloe humilis* var. *echinata* (Willd.) Baker 1896; **incl.***Aloe acuminata* var. *major* Salm-Dyck 1817; **incl.** *Aloe humilis* subvar. *semiguttata* Haw. 1821; **incl.** *Aloe subtuberculata* Haw. 1825 ≡ *Aloe humilis* var. *subtuberculata* (Haw.) Baker 1896; **incl.** *Aloe humilis* subvar. *minor* Salm-Dyck 1837; **incl.** *Aloe humilis* var. *candollei* Baker 1880.
Distribution: South Africa (Western Cape, Eastern Cape).

A. ibitiensis H.Perrier 1926, Mém. Soc. Linn. Normandie, Bot. 1(1): 30.
Incl. *Aloe ibityensis* hort. (*nom. nud.*, Art. 61.1).
Distribution: Madagascar (C).

A. imalotensis Reynolds 1957, J. South Afr. Bot. 23: 68.
Incl. *Aloe deltoideodonta* var. *contigua* H.Perrier 1926 ≡ *Aloe contigua* (H.Perrier) Reynolds 1958; **incl.** *Aloe deltoideodonta* forma *latiofolia* H.Perrier 1938 (*nom. nud.*, Art. 36.1); **incl.** *Aloe deltoideodonta* forma *longifolia* H.Perrier 1938 (*nom. nud.*, Art. 36.1); **incl.** *Aloe deltoideondonta* subforma *variegata* Boiteau ex H.Jacobsen 1954 (*nom. nud.*, Art. 36.1).
Distribution: Madagascar.

A. ×imerinensis Bosser 1968, Adansonia, n.s., 8(4): 510-512, ill..
Distribution: Madagascar.

A. immaculata Pillans 1934, South Afr. Gard. 24: 25.
Distribution: South Africa (Northern Prov.).

A. inamara L.C.Leach 1971, J. South Afr. Bot. 37(4): 259-266, ills..
Distribution: Angola.

A. inconspicua Plowes 1986, Aloe 23(2): 32-33, ills..
Distribution: South Africa (KwaZulu-Natal).

A. inermis Forssk. 1775, Fl. Aegypt.-Arab., 74.
Distribution: Yemen.

A. integra Reynolds 1936, Flow. Pl. South Afr. 16: t. 607 + text.
Distribution: Swaziland, South Africa (Mpumalanga).

A. inyangensis Christian 1936, Flow. Pl. South Afr. 16: t. 640 + text
Distribution: Zimbabwe.

A. inyangensis var. **inyangensis**
Distribution: Zimbabwe.

A. inyangensis var. **kimberleyana** S.Carter 1996, Kew Bull. 51(4): 777-779, ill..
Distribution: Zimbabwe.

A. isaloensis H.Perrier 1927, Bull. Acad. Malgache, n.s. 10: 20.
Distribution: Madagascar.

A. itremensis Reynolds 1955, J. South Afr. Bot. 22(1): 29-30, ills..
Distribution: Madagascar.

A. jacksonii Reynolds 1955, J. South Afr. Bot. 21(2): 59-61, t. 5.
Distribution: Ethiopia (Ogaden).

A. jucunda Reynolds 1953, J. South Afr. Bot. 19(1): 21-23, t. 11.
Distribution: Somalia.

A. juvenna Brandham & S.Carter 1979, Cact. Succ. J. Gr. Brit. 41(2): 27-29, ills..
Distribution: Kenya.

A. ×**keayi** Reynolds (*pro sp.*) 1963, J. South Afr. Bot. 29(2): 43-44, t. 6-7.

A. kedongensis Reynolds 1953, J. South Afr. Bot. 19(1): 4-6, t. 3-4.
  ≡ *Aloe nyeriensis* spp. *kedongensis* (Reynolds) S.Carter 1980.
Distribution: Kenya.

A. kefaensis M.G.Gilbert & Sebsebe 1997, Kew Bull. 52(1): 140-141.
Distribution: Ethiopia.

A. keithii Reynolds 1937, J. South Afr. Bot. 3(1): 47-49, t. 5.
Distribution: Swaziland.

A. ketabrowniorum L.E.Newton 1994, Brit. Cact. Succ. J. 12(2): 50-51, ills..
Distribution: Kenya.

A. khamiensis Pillans 1934, South Afr. Gard. 1934: 24.
Distribution: South Africa (Northern Cape).

A. kilifiensis Christian 1942, J. South Afr. Bot. 8(2): 169-170, t. 3.
Distribution: Kenya.

A. kniphofioides Baker 1890, Hooker's Icon. Pl. 1890: t. 1939.
  Incl. *Aloe marshallii* J.M.Wood & M.S.Evans 1897.
Distribution: South Africa (KwaZulu-Natal, Mpumalanga).

A. krapohliana Marloth 1908, Trans. Roy. Soc. South Afr. 1: 408.
Distribution: South Africa.

A. krapohliana var. **dumoulinii** Lavranos 1973, J. South Afr. Bot. 39(1): 41-43, ills..
Distribution: South Africa (Northern Cape).

A. krapohliana var. **krapohliana**
Distribution: South Africa (Northern Cape, Western Cape).

A. kraussi Baker 1880, J. Linn. Soc., Bot. 18: 159.
Distribution: South Africa (KwaZulu-Natal).

A. kulalensis L.E.Newton & Beentje 1990, Cact. Succ. J. (US) 62(5): 251-252, ills..
Distribution: Kenya (N).

A. labworana (Reynolds) S Carter 1994, Fl. Trop. East Afr., Aloaceae, 28.
  ≡ *Aloe schweinfurthii* var. *labworana* Reynolds 1956.
Distribution: Sudan (the), Uganda.

A. laeta A.Berger 1908, in Engler, A. (ed.), Pflanzenr. IV.38 (Heft 33): 256-257.
Distribution: Madagascar.

A. laeta var. **laeta**
Distribution: Madagascar.

A. laeta var. **maniaensis** H.Perrier 1926, Mém. Soc. Linn. Normandie, Bot. 1(1): 30.
Distribution: Madagascar.

A. lateritia Engl. 1895, Pfl.-welt Ost-Afr., Teil C, 140.
Distribution: Ethiopia, Kenya, Tanzania (United Republic of).

A. lateritia var. **graminicola** (Reynolds) S.Carter 1994, Fl. Trop. East Afr., Aloaceae, 17.
≡ *Aloe graminicola* Reynolds 1953; **incl.** *Aloe solaiana* Christian 1940.
Distribution: Ethiopia, Kenya.

A. lateritia var. **lateritia**
Incl. *Aloe boehmii* Engl. 1895; **incl.***Aloe campylosiphon* A.Berger 1904; **incl.** *Aloe amanensis* A.Berger 1905.
Distribution: Kenya, Tanzania (United Republic of).

A. lavranosii Reynolds 1964, J. South Afr. Bot. 30(4): 225-227, t. 31-32.
Distribution: Yemen.

A. leachii Reynolds 1965, J. South Afr. Bot. 31: 275.
Distribution: Tanzania (United Republic of).

A. leandrii Bosser 1968, Adansonia, n.s., 8(4): 505-507, ills..
Distribution: Madagascar.

A. leedalii S.Carter 1994, fl. Trop. East Afr., Aloaceae, 9-11, ills..
Distribution: Tanzania (United Republic of).

A. lensayuensis Lavranos & L.E.Newton 1976, Cact. Succ. J. (US) 48(6): 276-278, ills..
Distribution: Kenya.

A. lepida L.C.Leach 1974, J. South Afr. Bot. 40(2): 102-106, ills..
Distribution: Angola.

A. leptosiphon A.Berger 1905, Bot. Jahrb. Syst. 36: 66.
Incl. *Aloe greenwayi* Reynolds 1964.
Distribution: Tanzania (United Republic of).

A. lettyae Reynolds 1937, J. South Afr. Bot. 3: 137.
Distribution: South Africa (Northern Prov.).

A. lindenii Lavranos 1997, Cact. Succ. J. (US) 69(3): 149-151, ills..
Distribution: Somalia.

A. linearifolia A.Berger 1922, Bot. Jahrb. Syst. 57: 640.
Distribution: South Africa (KwaZulu-Natal).

A. lineata (Aiton) Haw. 1804, Trans. Linn. Soc. London 7: 18.
≡ *Aloe perfoliata* var. *lineata* Aiton 1789.
Distribution: South Africa.

A. lineata var. **lineata**
Incl. *Aloe lineata* var. *glaucescens* Haw. 1821; **incl.** *Aloe lineata* var. *viridis* Haw. 1821.
Distribution: South Africa (Western Cape, Eastern Cape).

A. lineata var. **muirii** (Marloth) Reynolds 1950, Aloes South Afr. [ed. 1], 205.
≡ *Aloe muirii* Marloth 1929.
Distribution: South Africa (Western Cape).

A. littoralis Baker 1878, Trans. Linn. Soc. London, Bot. 1: 263.
Incl. *Aloe rubrolutea* Schinz 1896; **incl.** *Aloe schinzii* Baker 1898.
Distribution: Angola, Botswana, Mozambique, Namibia, South Africa, Zambia, Zimbabwe.

A. lolwensis L.E.Newton 2001, Cact. Succ.J. (US) 73: 155-158, ills. -
Distribution: Kenya

A. lomatophylloides Balf.f. 1877, J. Linn. Soc., Bot. 16: 22.
≡ *Lomatophyllum lomatophylloides*(Balf.f.) Marais 1975.
Distribution: Mauritius (Rodrigues Island).

A. longistyla Baker 1880, J. Linn. Soc., Bot. 18: 158.
Distribution: South Africa (Western Cape, Eastern Cape).

A. luapulana L.C.Leach 1972, J. South Afr. Bot. 38(3): 185-188, ills..
Distribution: Zambia.

A. lucile-allorgeae Rauh 1998, Bradleya 16: 97-98, 100, ills..
Distribution: Madagascar.

A. luntii Baker 1894, Bull. Misc. Inform. [Kew] 1894: 342.
Distribution: Oman, Somalia, Yemen.

A. lutescens Groenew. 1938, Flow. Pl. South Afr. 18: t. 707 + text.
Distribution: South Africa (Northern Prov.).

A. macleayi Reynolds 1955, J. South Afr. Bot. 21(2): 55-57, t. 3-4.
Distribution: Sudan (the).

A. macra Haw. 1819, Suppl. Pl. Succ., 45, 105.
≡ *Phylloma macrum* (Haw.) Sweet 1827 ≡ *Lomatophyllum macrum* (Haw.)
Salm-Dyck *ex* Roem.& Schult. 1829.
Distribution: France (Réunion).

A. macrocarpa Tod. 1875, Hort. Bot. Panorm. 1: 36, t. 9.
Incl. *Aloe commutata* A. Engl. 1892; incl. *Aloe macrocarpa* var. *major* A.Berger 1908; **incl.**
Aloe edulis A.Chev. 1920; **incl.***Aloe barteri* Schnell 1953 (*nom. illeg.*, Art. 53.1).
Distribution: Benin, Cameroon, Djibouti, Eritrea, Ethiopia, Ghana, Mali, Nigeria, Sudan (the).

A. macroclada Baker 1883, J. Linn. Soc., Bot. 20: 273.
Distribution: Madagascar.

A. macrosiphon Baker 1898, in Thiselton-Dyer & al. Fl. Trop. Afr. 7: 459.
Incl. *Aloe mwanzana* Christian 1940; incl. *Aloe compacta* Reynolds 1961.
Distribution: Kenya, Rwanda, Tanzania (United Republic of), Uganda.

A. maculata All. 1773, Auct. Syn., 13.
Incl. *Aloe saponaria* var. *saponaria*; incl. *Aloe perfoliata* var. δ L. 1753; incl. *Aloe perfoliata*
var. θ L. 1753; **incl.** *Aloe perfoliata* var. λ L. 1753: **incl.** *Aloe disticha* Mill. 1768 (*nom. illeg.*,
Art. 53.1); incl. *Aloe maculosa* Lam. 1783; incl. *Aloe maculata* Medik. 1786 (*nom. illeg.*, Art.
53.1); incl. *Aloe perfoliata* var. *saponaria* Aiton 1789 ≡ *Aloe saponaria* (Aiton) Haw. 1804;
incl. *Aloe umbellata* DC. 1799; incl. *Aloe saponaria* var. *latifolia* Haw. 1804; incl. *Aloe
latifolia* Haw. 1812; **incl.***Aloe leptophylla* N.E.Br. *ex* Baker 1880; **incl.** *Aloe saponaria* var.
brachyphylla Baker 1880; **incl.** *Aloe leptophylla* var. *stenophylla* Baker 1896; **incl.** *Aloe
saponaria* var. *ficksburgensis* Reynolds 1937.
Distribution: Lesotho, South Africa (Eastern Cape, Free State, KwaZulu-Natal, Mpumalanga,
Western Cape), Swaziland.

A. madecassa H.Perrier 1926, Mém. Soc. Linn. Normandie, Bot. 1(1): 23.
Distribution: Madagascar.

A. madecassa var. **lutea** Guillaumin 1955, Bull. Mus. Nation. Hist. Nat., Sér. 2, 27: 86.
Distribution: Madagascar.

A. madecassa var. **madecassa**
Distribution: Madagascar.

A. marlothii A.Berger 1905, Bot. Jahrb. Syst. 38: 87.
Incl. *Aloe ferox* var. *xanthostachys* A.Berger 1908 (*incorrect name*, Art. 11.4); **incl.**
Aloe ferox A.Berger 1908 (*nom. illeg.*, Art. 53.1); **incl.** *Aloe marlothii* J.M.Wood 1912 (*nom. illeg.*, Art. 53.1).
Distribution: Botswana, Mozambique, South Africa, Swaziland.

A. marlothii ssp. **orientalis** Glen & D.S.Hardy 1987, Flow. Pl. Afr. 49(3-4): pl. 1943 + text.
Distribution: Mozambique, South Africa (KwaZulu-Natal).

A. marlothii var. **bicolor** Reynolds 1936, J. South Afr. Bot. 2(1): 34.
Distribution: South Africa.

A. marlothii var. **marlothii**
Incl. *Aloe supralaevis* var. *hanburyi* Baker 1896; **incl.** *Aloe spectabilis* Reynolds 1927.
Distribution: Botswana, Mozambique, South Africa (KwaZulu-Natal, Gauteng, Mpumalanga, Northern Prov., Northwest Prov.), Swaziland.

A. massawana Reynolds 1959, J. South Afr. Bot. 25: 207-109, pl. 18-19.
Incl. *Aloe kirkii* Baker 1894.
Distribution: Tanzania (United Republic of).

A. mawii Christian 1940, J. South Afr. Bot. 6(4): 186-188, t. 23.
Distribution: Malawi, Mozambique, Tanzania (United Republic of).

A. mayottensis A.Berger 1908, In Engler, A. (ed.), Pflanzenr. IV.38 (Heft 33): 246.
Distribution: Comoros (the).

A. mccoyi Lavranos & Mies 2001, Cact. Succ.J. (US) 73: 146-151, ills.
Distribution: Yemen.

A. mcloughlinii Christian 1951, Flow Pl. Afr. 28: t. 1112 + text.
Incl. *Aloe maclaughlinii* hort. (*nom. nud.*, Art. 61.1).
Distribution: Djibouti, Ethiopia.

A. medishiana Reynolds & P.R.O.Bally 1958, J. South Afr. Bot. 24(4): 186-187, t. 29.
Distribution: Somalia.

A. megalacantha Baker 1898, in Theselton-Dyer & al., Fl. Trop. Afr. 7: 469.
Distribution: Ethiopia, Somalia.

A. megalacantha ssp. **alticola** M.G.Gilbert & Sebsebe 1997, Kew Bull. 52(1): 150.
Distribution: Ethiopia.

A. megalacantha ssp. **megalacantha**
Incl. *Aloe magnidentata* I.Verd. & Christian 1947.
Distribution: Ethiopia, Somalia.

A. megalocarpa Lavranos 1998, Kakt. and Sukk. 49(7): 162-163, ills..
Distribution: Madagascar (N).

A. melanacantha A.Berger 1905, Bot. Jahrb. Syst. 36: 63.
Distribution: Namibia, South Africa.

A. melanacantha var. **erinacea** (D.S.Hardy) G.D.Rowley 1980, Excelsa 9: 71, 80
≡ *Aloe erinacea* D.S.Hardy 1971.
Distribution: Namibia.

A. melanacantha var. **melanacantha**
Distribution: Namibia, South Africa (Northern Cape).

A. menachensis (Schweinf.) Blatt. 1936, Fl. Arab., 463.
≡ *Aloe percrassa* var. *menachensis* Schweinf. 1894 ≡ *Aloe trichosantha* var. *menachensis* (Schweinf.) A.Berger 1908.
Distribution: Yemen.

A. mendesii Ryenolds 1964, J. South Afr. Bot. 30(1): 31-32, t. 10.
Distribution: Angola, Namibia.

A. menyharthii Baker 1898, in Theselton-Dyer & al., Fl. Trop. Afr. 7: 459.
Distribution: Malawi, Mozambique.

A. menyharthii ssp. **ensifolia** S.Carter 1996, Kew Bull. 51(4): 783-784.
Distribution: Mozambique.

A. menyharthii ssp. **menyharthii**
Distribution: Malawi, Mozambique.

A. metallica Engl. & Gilg 1903, in Warburg, Kunene-Sambesi Exped., 191.
Distribution: Angola.

A. meyeri Van Jaarsv. 1981, J. South Afr. Bot. 47(3): 567-571, ills..
Incl. *Aloe richtersveldensis* Venter & Beukes 1982.
Distribution: Namibia, South Africa (Northern Cape).

A. micracantha Haw. 1819, Suppl. Pl. Succ., 105.
Incl. *Aloe micracantha* Link & Otto 1825 (*nom. illeg.*, Art. 53.1).
Distribution: South Africa (Eastern Cape).

A. microdonta Chiov. 1928, Pl. Nov. Min. Not. Ethiop. 1: 7.
Distribution: Kenya, Somalia.

A. microstigma Salm-Dyck 1854, Monogr. Gen. Aloes & Mesembr. Sect. 26: fig. 4 (fasc. 6: fig. 11).
Incl. *Aloe juttae* Dinter 1923; incl. *Aloe brunnthaleri* A.Berger *ex* Cammerl. 1933.
Distribution: South Africa (Western Cape, Eastern Cape).

A. millotii Reynolds 1956, J. South Afr. Bot. 22(1): 23-26, ills..
Distribution: Madagascar (Toliara).

A. milne-redheadii Christian 1940, J. South Afr. Bot. 6(4): 177-179, t. 18.
Distribution: Angola, Zambia.

A. minima Baker 1895, Hooker's Icon. Pl. 25: t. 2423 + text.
≡ *Leptaloe minima* (Baker) Stapf 1933.
Distribution: South Africa, Swaziland.

A. minima var. **blyderivierensis** (Groenew.) Reynolds 1947, J. South Afr. Bot. 13(2): 101, t. 15: fig. 2.
≡ *Leptaloe blyderivierensis* Groenew. 1938.
Distribution: South Africa (Mpumalanga).

A. minima var. **minima**
   Incl. *Aloe parviflora* Baker 1901.
   Distribution: South Africa (KwaZulu-Natal), Swaziland.

A. mitriformis Mill. 1768, Gard. Dict., Ed. 8, no. 1.
   Incl. *Aloe mitriformis* var. *humilior* Haw.; incl. *Aloe perfoliata* var. v L. 1753; incl. *Aloe perfoliata* var. *mitriformis* Aiton 1789; incl. *Aloe perfoliata* var. ξ Willd. 1799; incl. *Aloe mitriformis* var. *elatior* Haw. 1804; incl. *Aloe xanthacantha* Salm-Dyck 1854; incl. *Aloe parvispina* Schönland 1905.
   Distribution: South Africa (Western Cape).

A. modesta Reynolds 1956, J. South Afr. Bot. 22(2): 85-86, ills..
   Distribution: South Africa (Mpumalanga).

A. molederana Lavranos & Glen 1989, Flow. Pl. Afr. 50(2): t. 1982 + 6 pp. of text, diags., ill..
   Distribution: Somalia.

A. monotropa I.Verd. 1961, Flow. Pl. Afr. 34: t. 1342 + text.
   Distribution: South Africa (Northern Prov.).

A. monticola Reynolds 1957, J. South Afr. Bot. 23(1): 7-9, t. 7-8.
   Distribution: Ethiopia (Tigre).

A. morijensis S.Carter & Brandham 1979, Cact. Succ. J. Gr. Brit. 41(1): 3-4, ills..
   Distribution: Kenya, Tanzania (United Republic of).

A. mubendiensis Christian 1942, J. South Afr. Bot. 8(2): 172-173, t. 5.
   Distribution: Uganda.

A. mudenensis Reynolds 1937, J. South Afr. Bot. 3(1): 39-42, t. 1.
   Distribution: South Africa (KwaZulu-Natal).

A. multicolor L.E.Newton 1994, Brit. Cact. Succ. J. 12(2): 51-52, ills..
   Distribution: Kenya.

A. munchii Christian 1951, Flow. Pl. Afr. 28: t. 1091 + text.
   Distribution: Mozambique, Zimbabwe.

A. murina L.E.Newton 1992, Taxon 41(1): 31-33, ills..
   Distribution: Kenya.

A. musapana Reynolds 1964, J. South Afr. Bot. 30(3): 125-126, t. 22.
   Distribution: Zimbabwe.

A. mutabilis Pillans 1933, South Afr. Gard. 23: 168.
   Distribution: South Africa (Gauteng, Mpumalanga, Northern prov., North-West Prov.)

A. myriacantha (Haw.) Schult. & Schult.f. 1829, Syst. Veg. 7(1): 704.
   ≡ *Bowiea myriacantha* Haw. 1827 ≡ *Leptaloe myriacantha* (Haw.) Stapf 1933; incl. *Aloe johnstonii* Baker 1887; incl. *Aloe caricina* A.Berger 1905; incl. *Aloe graminifolia* A.Berger 1905.
   Distribution: Burundi, Democratic Republic of the Congo (the), Kenya, Malawi, Rwanda, South Africa (Eastern Cape, KwaZulu-Natal), Tanzania (United Republic of), Uganda, Zimbabwe.

A. mzimbana Christian 1941, Flow. Pl. South Afr. 21: t. 838 + text.
   Distribution: Democratic Republic of the Congo (the), Malawi, Tanzania (United Republic of), Zambia.

A. namibensis Giess 1970, Mitt. Bot. Staatssamml. München 8: 123-126.
Distribution: Namibia.

A. namorokaensis (Rauh) L.E.Newton & G.D.Rowley 1998, Bradleya 16: 114.
≡ *Lomatophyllum namorokaense* Rauh 1998.
Distribution: Madagascar (W).

A. ngongensis Christian 1942, J. South Afr. Bot. 8(2): 170-172, t. 4.
Distribution: Kenya, Tanzania (United Republic of).

A. niebuhriana Lavranos 1965, J. South Afr. Bot. 31(1): 68-71, t. 13.
Distribution: Yemen.

A. nubigena Groenew. 1936, Tydskr. Wetensk. Kuns 14: 135-137.
Distribution: South Africa (Mpumalanga).

A. nuttii Baker 1897, Hooker's Icon. Pl. 1897: t. 2513 + text.
Incl. *Aloe brunneo-punctata* Engl. & Gilg 1903; **incl.** *Aloe corbisieri* De Wild. 1921; **incl.** *Aloe mketiensis* Christian 1940.
Distribution: Angola, Democratic Republic of the Congo (the), Malawi, Tanzania (United Republic of), Zambia,.

A. nyeriensis Christian *ex* I.Verd. 1952, Flow. Pl. Afr. 29: t. 1126 + text.
Incl. *Aloe ngobitensis* Reynolds 1953.
Distribution: Kenya.

A. occidentalis (H.Perrier) L.E.Newton & G.D.Rowley 1997, Excelsa 17: 61.
≡ *Lomatophyllum occidentale* H.Perrier 1926.
Distribution: Madagascar (W).

A. officinalis Forssk. 1775, Fl. Aegypt.-Arab, 73.
≡ *Aloe vera* var. *officinalis* (Forssk.) Baker 1880; **incl.** *Aloe maculata* Forssk. 1775 (*nom illeg.*, Art. 53.1); **incl.** *Aloe vera* var. *angustifolia* Schweinf. 1894 ≡ *Aloe officinalis* var. *angustifolia* (Schweinf.) Lavranos 1965.
Distribution: Saudi Arabia, Yemen.

A. oligophylla Baker 1883, J. Linn. Soc., Bot. 20: 272.
≡ *Lomatophyllum oligophyllum* (Baker) H.Perrier 1926.
Distribution: Madagascar.

A. orientalis (H.Perrier) L.E.Newton & G.D.Rowley 1997, Excelsa 17: 61.
≡ *Lomatophyllum orientale* H.Perrier 1926.
Distribution: Madagascar.

A. ortholopha Christian & Milne-Redh. 1933, Bull. Misc. Inform [Kew] 1933: 478.
Distribution: Zimbabwe.

A. otallensis Baker 1898, in Thiselton-Dyer & al., Fl. Trop. Afr. 7: 458.
Incl. *Aloe boranensis* Cufodontis 1939.
Distribution: Ethiopia.

A. pachygaster Dinter 1924, Repert. Spec. Nov. Regni Veg. 19: 179.
Distribution: Namibia.

A. paedogona A.Berger 1906, J. Bot. 44: 57.
Distribution: Angola, Namibia.

A. palmiformis Baker 1878, Trans. Linn. Soc. London, Bot. 1: 263.
Distribution: Angola.

A. parallelifolia H.Perrier 1926, Mém. Soc. Linn. Normandie, Bot. 1(1): 31.
Distribution: Madagascar.

A. parvibracteata Schönland 1907, Rec. Albany Mus. 2: 139.
Incl. *Aloe komatiensis* Reynolds 1936; **incl.** *Aloe pongolensis* Reynolds 1936; **incl.** *Aloe decurvidens* Groenew. 1937; **incl.** *Aloe lusitanica* Groenew. 1937; **incl.** *Aloe pongolensis* var. *zuluensis* Reynolds 1937 ≡ *Aloe parvibracteata* var. *zuluensis* (Reynolds) Reynolds 1950.
Distribution: Mozambique, South Africa (KwaZulu-Natal), Mpumalanga), Swaziland.

A. parvicapsula Lavranos & Collen. 2000, in Cact. Succ. J. (USA), 72(2): 84
Distribution: Saudi Arabia

A. parvicoma Lavranos & Collen. 2000, in Cact. Succ. J. (USA), 72(1): 21
Distribution: Saudi Arabia

A. parvidens M.G.Gilbert & Sebsebe 1992, Kew Bull. 47(4): 650.
Distribution: Ethiopia, Kenya, Somalia, Tanzania (United Republic of).

A. parvula A.Berger 1908, In Engler, A. (ed.), Pflanzenr. IV.38 (Heft 33): 172-173.
≡ *Lemeea parvula* (A.Berger) P.V.Heath 1994; **incl.** *Aloe sempervivoides* H.Perrier 1926.
Distribution: Madagascar.

A. patersonii B.Mathew 1978, Kew Bull. 32(2): 321-322.
Distribution: Democratic Republic of the Congo (the).

A. pearsonii Schönland 1911, Rec. Albany Mus. 2: 229.
Distribution: Namibia, RSA (Northern Cape).

A. peckii P.R.O.Bally & I.Verd. 1956, Flow. Pl. Afr. 31: t. 1214 + text.
Distribution: Somalia.

A. peglerae Schönland 1904, Rec. Albany Mus. 1: 120.
Distribution: South Africa (Gautang).

A. pembana L.E.Newton 1998, Cact. Succ. J. (US) 70(1): 27-31, ills..
≡ *Lomatophyllum pembanum* (L.E.Newton) Rauh 1998.
Distribution: Tanzania (United Republic of) (Pemba).

A. pendens Forssk. 1775, Fl. Aegypt.-Arab., 74.
Incl. *Aloe variegata* Forssk. 1775 (*nom. illeg.*, Art. 53.1); **incl.** *Aloe arabica* Lam. 1783; **incl.** *Aloe dependens* Steud. 1840.
Distribution: Yemen.

A. penduliflora Baker 1888, Gard. Chron., ser. 3, 4: 178.
Distribution: Kenya.

A. percrassa Tod. 1875, Hort. Bot. Panorm. 1: 81, t. 21.
≡ *Aloe abyssinica* var. *percrassa* (Tod.) Baker 1880; **incl.** *Aloe schimperi* Schweinf. 1894 (*nom. illeg.*, Art. 53.1); **incl.** *Aloe oligospila* Baker 1902; **incl.** *Aloe schimperi* G.Karst. & Schenk 1905 (*nom. illeg.*, Art. 53.1).
Distribution: Eritrea, Ethiopia.

A. perrieri Reynolds 1956, J. South Afr. Bot. 22(3): 131.
Incl. *Aloe parvula* H.Perrier 1926 (*nom. illeg.*, Art. 53.1).
Distribution: Madagascar.

A. perryi Baker 1881, J. Linn. Soc., Bot. 18: 161.
Distribution: Yemen (Socotra).

A. petricola Pole-Evans 1917, Trans. Roy. Soc. South Afr. 5: 707.
Distribution: South Africa (Mpumalanga).

A. petrophila Pillans 1933, South Afr. Gard. 23: 213.
Distribution: South Africa (Northern Prov.).

A. peyrierasii Cremers 1976, Adansonia, n.s., 15(4): 500-503, ill..
≡ *Lomatophyllum peyrierasii* (Cremers) Rauh 1998.
Distribution: Madagascar.

A. pictifolia D.S.Hardy 1976, Bothalia 12(1): 62-64, ills..
Distribution South Africa (Eastern Cape).

A. pillansii L.Guthrie 1928, J. Bot. 66: 15.
Distribution: Namibia, South Africa (Northern Cape).

A. pirottae A.Berger 1905, Bot. Jahrb. Syst. 36: 65.
Distribution: Ethiopia, Kenya.

A. plicatilis (L.) Mill. 1768, Gard. Dict., Ed. 8, no. 7.
≡ *Aloe disticha* var. *plicatilis* L. 1753 ≡ *Rhipidodendrum plicatile* (L.) Haw. 1821; **incl.** *Aloe linguaeformis* L.f. 1782; **incl.** *Aloe tripetala* Medik. 1783; **incl.** *Aloe lingua* Thunb. 1785; **incl.** *Kumara disticha* Medik. 1786 ≡ *Rhipidodendrum distichum* (Medik.) Willd. 1811; **incl.** *Aloe flabelliformis* Salisb. 1796; **incl.** *Aloe plicatilis* var. *major* Salm-Dyck 1817.
Distribution: South Africa (Western Cape).

A. plowesii Reynolds 1964, J. South Afr. Bot. 30(2): 71-73, t. 14.
Distribution: Mozambique.

A. pluridens Haw. 1824, Philos. Mag. J. 64: 299.
Incl. *Aloe atherstonei* Baker 1880; **incl.** *Aloe pluridens* var. *beckeri* Schönland 1903.
Distribution: South Africa (Eastern Cape, KwaZulu-Natal).

A. polyphlla Schönland *ex* Pillans 1934, South Afr. Gard. 24: 267.
Distribution: Lesotho.

A. porphyrostachys Lavranos & Collen. 2000, in Cact. Succ. J. (USA), 72(1): 18
Distribution: Saudi Arabia

A. powysiorum L.E.Newton & Beentje 1990, Cact. Succ. J. (US) 62(5): 252-255, ills..
Distribution: Kenya.

A. pratensis Baker 1880, J. Linn. Soc., Bot. 18: 156.
Distribution: Lesotho, South Africa (Eastern Cape, KwaZulu-Natal).

A. pretoriensis Pole-Evans 1915, Gard. Chron., ser. 3, 56: 106.
Distribution: South Africa (Gauteng, Mpumalanga, Northern Prov.), Swaziland, Zimbabwe.

A. prinslooi I.Verd. & D.S.Hardy 1965, Flow. Pl. Afr. 37: t. 1453 + text.
Distribution: South Africa (KwaZulu-Natal).

A. procera L.C.Leach 1974, J. South Afr. Bot. 40(2): 117-121, ills..
Distribution: Angola.

A. propagulifera (Rauh & Razaf.) L.E.Newton & G.D.Rowley 1998, Bradleya 16: 114.
 ≡ *Lomatophyllum propaguliferum* Rauh & Razaf. 1998.
 Distribution: Madagascar (C-E).

A. prostrata (H.Perrier) L.E.Newton & G.D.Rowley 1997, Excelsa 17: 61.
 ≡ *Lomatophyllum prostratum* H.Perrier 1926.
 Distribution: Madagascar.

A. prostrata ssp. **pallida** Rauh & Mangelsdorff
 Distribution: Madagascar

A. pruinosa Reynolds 1936, J. South Afr. Bot. 2(2): 122-124, t. 17.
 Distribution: South Africa (KwaZulu-Natal).

A. pseudorubroviolacea Lavranos & Collen.2000, in Cact. Succ. J. (USA), 72(1): 17
 Distribution: Saudi Arabia

A. pubescens Reynolds 1957, J. South Afr. Bot. 23(1): 10-12 , t. 10-11.
 Distribution: Ethiopia.

A. pulcherrima M.G.Gilbert & Sebsebe 1997, Kew Bull. 52(1): 147-148.
 Distribution: Ethiopia.

A. purpurea Lam. 1783, Encycl. 1: 85.
 ≡ *Lomatophyllum purpureum* (Lam.) T.Durand & Schinz 1895; **incl.** *Dracaena marginata*
 Aiton 1789 (*nom. illeg.*, Art. 52) ≡ *Aloe marginata* (Aiton) Willd. 1809 (*nom. illeg.*, Art.
 53.1); **incl.** *Aloe marginalis* A.DC. 1800 (*nom. illeg.*, Art. 52); **incl.** *Dracaena dentata* Pers.
 1805 (*nom. illeg.*, Art. 52); **incl.** *Lomatophyllum borbonicum* Willd. 1811 (*nom. illeg.*, Art.
 52.1); **incl.** *Phylloma aloiflorum* Ker Gawl. 1813 (*nom. illeg.*, Art. 52.1) ≡ *Lomatophyllum*
 aloiflorum (Ker Gawl.) G.Nicholson 1885; **incl.** *Aloe rufocincta* Haw. 1819 ≡ *Phylloma*
 rufocinctum (Haw.) Sweet 1827 ≡ *Lomatophyllum rufocinctum* (Haw.) Salm-Dyck *ex* Roem.&
 Schult. 1829; **incl.** *Lomatophyllum marginata* Hoffmanns. 1824 (*nom. nud.*, Art. 32.1c).
 Distribution: Mauritius.

A. pustuligemma L.E.Newton 1994, Brit. Cact. Succ. J. 12(2): 53-54, ills..
 Distribution: Kenya.

A. ×qaharensis Lavranos & Collen. 2000, in Cact. Succ. J. (USA), 72(2): 87
 Distribution: Saudi Arabia

A. rabaiensis Rendle 1895, J. Linn. Soc., Bot. 30: 410.
 Distribution: Kenya, Somalia, Tanzania (United Republic of).

A. ramosissima Pillans 1937, J. South Afr. Bot. 5(3): 66-67, t. 7.
 Distribution: Namibia, South Africa (Northern Cape).

A. rauhii Reynolds 1963, J. South Afr. Bot. 29(4): 151-152, t. 24-25.
 ≡ *Guillauminia rauhii* (Reynolds) P.V.Heath 1994.
 Distribution: Madgascar.

A. reitzii Reynolds 1937, J. South Afr. Bot. 3(3): 135-137, t. 20.
 Distribution: South Africa.

A. reitzii var. **reitzii**
 Distribution: South Africa (Mpumalanga).

A. reitzii var. **vernalis** D.S.Hardy 1981, Bothalia 13(3/4): 451-452, ills..
Distribution: South Africa (KwaZulu-Natal).

A. retrospiciens Reynolds & P.R.O.Bally 1958, J. South Afr. Bot. 24(4): 182-184, t. 25-26.
Incl. *Aloe ruspoliana* var. *dracaeniformis* A.Berger 1908.
Distribution: Somalia.

A. reynoldsii Letty 1934, Flow. Pl. South Afr. 14: t. 558 + text.
Distribution: South Africa (Eastern Cape).

A. rhodesiana Rendle 1911, J. Linn. Soc., Bot. 40: 215.
Incl. *Aloe eylesii* Christian 1936.
Distribution: Mozambique, Zimbabwe.

A. richardsiae Reynolds 1964, J. South Afr. Bot. 30(2): 67-69, t. 12-13.
Distribution: Tanzania (United Republic of).

A. rigens Reynolds & P.R.O.Bally 1958, J. South Afr. Bot. 24(4): 177-179, t. 20-21.
Distribution: Somalia, Yemen.

A. rigens var. **mortimeri** Lavranos 1967, Cact. Succ. J. (US) 39(4): 123-125, ills..
Distribution: Yemen.

A. rigens var. **rigens**
Distribution: Somalia.

A. rivae Baker 1898, in Thiselton-Dyer & al., Fl. Trop. Afr. 7: 465.
Distribution: Ethiopia, Kenya.

A. rivierei Lavranos & L.E.Newton 1977, Cact. Succ. J. (US) 49(3): 114-116, ills..
Distribution: Yemen.

A. rosea (H.Perrier) L.E.Newton & G.D.Rowley 1997, Excelsa 17: 61.
≡ *Lomatophyllum roseum* H.Perrier 1926.
Distribution: Madagascar.

A. rubroviolacea Schweinf. 1894, Bull. Herb. Boissier 2 (App.2): 71.
Distribution: Saudi Arabia, Yemen.

A. ruffingiana Rauh & Petignat 1999 in Kakt. And. Sukk. 50(11): 271
Distribution: Madagascar

A. rugosifolia M.G.Gilbert & Sebsebe 1992, Kew Bull. 47(4): 652-653.
Incl. *Aloe otallensis* var. *elongata* A.Berger 1908.
Distribution: Ethiopia, Kenya.

A. rupestris Baker 1896, in Thiselton-Dyer, Fl. Cap. 6: 327-328.
Incl. *Aloe pycnacantha* MacOwan ms. (*nom. nud.*, Art. 29.1); **incl.** *Aloe nitens* Baker 1880
(*nom. illeg.*, Art. 53.1).
Distribution: Mozambique, South Africa (KwaZulu-Natal), Swaziland.

A. rupicola Reynolds 1960, J. South Afr. Bot. 26(2): 89-91, t. 10-11.
Distribution: Angola.

A. ruspoliana Baker 1898, in Thiselton-Dyer & al., Fl. Trop. Afr. 7: 460.
Incl. *Aloe stephaninii* Chiov. 1916; **incl.** *Aloe jex-blakeae* Christian 1942.
Distribution: Ethiopia, Kenya, Somalia.

A. sabaea Schweinf. 1894, Bull. Herb. Boissier 2(App. 2): 74.
Incl. *Aloe gillilandii* Reynolds 1962.
Distribution: Saudi Arabia, Yemen.

A. saundersiae (Reynolds) Reynolds 1947, J. South Afr. Bot. 13(2): 103, ills..
≡ *Leptaloe saundersiae* Reynolds 1936; incl. *Aloe minima* Medley-Wood 1906 (*nom. illeg.*, Art. 53.1).
Distribution: South Africa (KwaZulu-Natal).

A. scabrifolia L.E.Newton & Lavranos 1990, Cact. Succ. J. (US) 62(5): 219-221, ills..
Distribution: Kenya.

A. schelpei Reynolds 1961, J. South Afr. Bot. 27(1): 1-3, t. 102.
Distribution: Ethiopia.

A. schilliana L.E.Newton & G.D.Rowley 1997, Excelsa 17: 61.
Incl. *Lomatophyllum viviparum* H.Perrier 1926.
Distribution: Madagascar.

A. schoelleri Schweinf. 1894, Bull. Herb. Boissier 2(App. 2): 107.
Distribution: Eritrea.

A. schomeri Rauh 1966, Kakt. and Sukk. 17(2): 22-24, ills..
Distribution: Madagascar (S).

A. schweinfurthii Baker 1880, J. Linn. Soc., Bot. 18: 175.
Incl. *Aloe barteri* var. *lutea* A.Chev. 1913; incl. *Aloe trivialis* A.Chev. 1952 (*nom. nud.*, Art. 36.1).
Distribution: Benin, Burkina Faso, Democratic Republic of the Congo (the), Ghana, Mali, Nigeria, Sudan (the), Uganda.

A. scobinifolia Reynolds & P.R.O.Bally 1958, J. South Afr. Bot. 24(4): 174-175, t. 17-18.
Distribution: Somalia.

A. scorpioides L.C.Leach 1974, J. South Afr. Bot. 40(2): 106-111, ills..
Distribution: Angola.

A. secundiflora Engl. 1895, Pfl.-welt Ost-Afr., Teil C, 140.
Distribution: Ethiopia, Kenya, Rwanda, Tanzania (United Republic of).

A. secundiflora var. **secundiflora**
Incl. *Aloe engleri* A.Berger 1905; incl. *Aloe floramaculata* Christian 1940; incl. *Aloe marsabitensis* I.Verd. & Christian 1940.
Distribution: Ethiopia, Kenya, Rwanda, Tanzania (United Republic of).

A. secundiflora var. **sobolifera** S.Carter 1994, Fl. Trop. East Afr., Aloaceae, 32-33, ills..
Distribution: Tanzania (United Republic of).

A. seretii De Wild. 1921, Pl. Bequaert. 1: 28.
Distribution: Democratic Republic of the Congo (the).

A. serriyensis Lavranos 1965, J. South Afr. Bot. 31(1): 76-77, t. 15.
Distribution: Yemen.

A. shadensis Lavranos & Collen. 2000, in Cact. Succ. J. (USA), 72(2): 82
Distribution: Saudi Arabia

A. sheilae Lavranos 1985, Cact. Succ. J. (US) 57(2): 71-72, ills..
Distribution: Saudi Arabia.

A. silicicola H.Perrier 1926, Mém. Soc. Linn. Normandie, Bot. 1(1): 42.
Distribution: Madagascar.

A. simii Pole-Evans 1917, Trans. Roy. Soc. South Afr. 5: 704.
Distribution: South Africa (Mpumalanga).

A. sinana Reynolds 1957, J. South Afr. Bot. 23(1): 3-5, t. 3-4.
Distribution: Ethiopia.

A. sinkatana Reynolds 1957, J. South Afr. Bot. 23(2): 39-42, t. 14-16.
Distribution: Sudan (the).

A. sladeniana Pole-Evans 1920, Ann. Bolus Herb. 3(1): 13.
Incl. *Aloe carowii* Reynolds 1938.
Distribution: Namibia.

A. socialis (H.Perrier) L.E.Newton & G. D. Rowley 1997, Excelsa 17: 61.
≡ *Lomatophyllum sociale* H.Perrier 1926.
Distribution: Madagascar.

A. somaliensis W.Watson 1899, Gard. Chron., ser. 3, 26: 430.
Distribution: Somalia.

A. somaliensis var. **marmorata** Reynolds & P.R.O.Bally 1964, J. South Afr. Bot. 30(4): 222-223, t. 30.
Distribution: Somalia.

A. somaliensis var. **somaliensis**
Distribution: Somalia.

A. soutpansbergensis I.Verd. 1962, Flow. Pl. Afr. 35: t. 1381 + text.
Distribution: South Africa (Northern Prov.).

A. speciosa Baker 1880, J. Linn. Soc., Bot. 18: 178.
Distribution: South Africa (Western Cape, Eastern Cape).

A. spicata L.f. 1782, Suppl. Pl., 205.
Incl. *Aloe sessiliflora* Pole-Evans 1917; incl. *Aloe tauri* L.C.Leach 1968.
Distribution: Mozambique, South Africa (KwaZulu-Natal, Mpumalanga), Swaziland, Zimbabwe.

A. splendens Lavranos 1965, J. South Afr. Bot. 31(1): 77-80, t. 16.
Distribution: Yemen.

A. squarrosa Baker 1883, Proc. Roy. Soc. Edinburgh 12: 97.
Incl. *Aloe concinna* Baker 1898 (*nom. illeg.*, Art. 53.1); incl. *Aloe zanzibarica* Milne-Redh. 1947.
Distribution: Yemen (Socotra).

A. steffanieana Rauh 2000, in Kakt. And. Sukk., 51(3): 73, 72
Distribution: Madagascar

A. steudneri Schweinf. 1894, Bull. Herb. Boissier 2(App.2): 73.
Distribution: Eritrea, Ethiopia.

A. striata Haw. 1804, Trans. Linn. Soc. London 7: 18
Distribution: Namibia, South Africa.

A. striata ssp. **karasbergensis** (Pillans) Glen & D.S.Hardy 1987, South Afr. J. Bot. 53(6): 491.
≡ *Aloe karasbergensis* Pillans 1928.
Distribution: Namibia, South Africa (Northern Cape).

A. striata ssp. **komaggasensis** (Kritzinger & Van Jaarsv.) Glen & D.S.Hardy 1987, South Afr. J. Bot. 53(6): 491.
≡ *Aloe komaggasensis* Kritzinger & Van Jaarsv. 1985.
Distribution: South Africa (Northern Cape).

A. striata ssp. **striata**
Incl. *Aloe rhodocincta* hort. *ex* Baker; incl. *Aloe paniculata* Jacq. 1809; incl. *Aloe albocincta* Haw. 1819; incl. *Aloe hanburyana* Naudin 1875; incl. *Aloe striata* var. *oligospila* Baker 1894.
Distribution: South Africa (Western Cape, Eastern Cape).

A. striatula Haw. 1825, Philos. Mag. J. 1825: 281.
Distribution: Lesotho, South Africa.

A. striatula var. **caesia** Reynolds 1936, Flow. Pl. South Afr. 16: t. 633 + text.
Incl. *Aloe striatula* forma *typica* Resende (*nom. nud.*, Art. 24.3); incl. *Aloe striatula* forma *conimbricensis* Resende 1943; incl.*Aloe stratula* forma *haworthii* Resende 1943.
Distribution: South Africa (Eastern Cape).

A. striatula var. **striatula**
Incl. *Aloe macowanii* Baker 1880; incl. *Aloe aurantiaca* Baker 1892; incl. *Aloe cascadensis* Kuntze 1898.
Distribution: Lesotho, South Africa (Eastern Cape).

A. suarezensis H.Perrier 1926, Mém. Soc. Linn. Normandie, Bot. 1(1): 21.
Distribution: Madagascar.

A. subacutissima G.D.Rowley 1973, Nation. Cact. Succ. J. 28(1): 6.
≡ *Aloe deltoideodonta* var. *intermedia* H.Perrier 1926 ≡ *Aloe intermedia* (H.Perrier) Reynolds 1957 (*nom. illeg.*, Art. 53.1).
Distribution: Madagascar.

A. succotrina All. 1773, Auct. Syn., 13.
Incl. *Aloe perforliata* var. ξ L. 1753; incl. *Aloe soccotrina* Garsault 1767 (*nom. nud.*, Art. 32.8); incl. *Aloe vera* Mill. 1768 (*nom. illeg.*, Art. 53.1); incl. *Aloe succotrina* Lam. 1783 (*nom. illeg.* Art. 53.1); incl. *Aloe perfoliata* var. *purpurascens* Aiton 1789 ≡ *Aloe purpurascens* (Aiton) Haw. 1804; incl. *Aloe perfoliata* var. *succotrina* Aiton 1789; incl. *Aloe sinuata* Thunb. 1794; incl. *Aloe soccotrina* var. *purpurascens* Ker Gawl. 1812 (*nom. nud.*, Art. 43.1); incl. *Aloe soccotrina* Schult. & Schult.f. 1829 (*nom. nud.*, Art. 61.1); incl. *Aloe succotrina* var. *saxigena* A.Berger 1908.
Distribution: South Africa (Western Cape).

A. suffulta Reynolds 1937, J. South Afr. Bot. 3: 151.
Incl. *Aloe subfulta* hort. (*nom. nud.*, Art. 61.1).
Distribution: Mozambique, South Africa (KwaZulu-Natal).

A. suprafoliata Pole-Evans 1916, Trans. Roy. Soc. South Afr. 5: 603.
Incl. *Aloe suprafoliolata* hort. (*nom. nud.*, Art. 61.1).
Distribution: South Africa (KwaZulu-Natal, Mpumalanga), Swaziland.

A. suzannae Decary 1921, Bull. Econ. Madag. 18: 26.
Distribution: Madagascar.

A. swynnertonii Rendle 1911, J. Linn. Soc., Bot. 40: 215.
Incl. *Aloe chimanimaniensis* Christian 1936; incl. *Aloe melsetterensis* Christian 1938.
Distribution: South Africa (Northern Prov.), Zimbabwe.

A. tenuior Haw. 1825, Philos. Mag. J. 1825: 281.
Incl. *Aloe tenuior* var. *glaucescens* Zahlbr. 1900; incl. *Aloe tenuior* var. *decidua* Reynolds
1936; incl. *Aloe tenuior* var. *rubriflora* Reynolds 1936; incl. *Aloe tenuior* var. *densiflora*
Reynolds 1950.
Distribution: South Africa (Eastern Cape).

A. tewoldei M.G.Gilbert & Sebsebe 1997, Kew Bull. 52(1): 143.
Distribution: Ethiopia.

A. thompsoniae Groenew. 1936, Tydskr. Wetensk. Kuns 14: 64.
Distribution: South Africa (Northern Prov.).

A. thorncroftii Pole-Evans 1917, Trans. Roy. Soc. South Afr. 5: 709.
Distribution: South Africa (Mpumalanga).

A. thraskii Baker 1880, J. Linn. Soc., Bot. 18: 180.
Incl. *Aloe candelabrum* Engl. & Drude 1910 (*nom. illeg.*, Art. 53.1).
Distribution: South Africa (Eastern Cape, KwaZulu-Natal).

A. tomentosa Deflers 1889, Voy. Yemen 211.
Incl. *Aloe tomentosa* forma *viridiflora* Lodé 1997 (*nom. nud.*, Art. 34.1b, 36.1).
Distribution: Saudi Arabia, Yemen.

A. tormentorii (Marais) L.E.Newton & G.D.Rowley 1997, Excelsa 17: 61.
≡ *Lomatophyllum tormentorii* Marais 1975.
Distribution: Mauritius.

A. tororoana Reynolds 1953, Flow. Pl. Afr. 29: t. 1144 + text.
Distribution: Uganda.

A. torrei I.Verd. & Christian 1946, Flow. Pl. Afr. 25: t. 987 + text.
Distribution: Mozambique.

A. trachyticola (H.Perrier) Reynolds 1957, J. South Afr. Bot. 23(2): 72-73, t. 26-27.
≡ *Aloe capitata* var. *trachyticola* H.Perrier 1926.
Distribution: Madagascar.

A. trichosantha A.Berger 1905, Bot. Jahrb. Syst. 36: 62.
Distribution : Eritrea, Ethiopia.

A. trichosantha ssp. **longiflora** M.G.Gilbert & Sebsebe 1997, Kew Bull. 52(1): 142-143.
Distribution: Ethiopia.

A. trichosantha ssp. **trichosantha**
Incl. *Aloe percrassa* Schweinf. 1894 (*nom. illeg.*, Art. 53.1); incl. *Aloe percrassa* var. *albo-
picta* Schweinf. 1894 (*incorrect name*, Art. 11.4).
Distribution: Eritrea, Ethiopia.

A. trigonantha L.C.Leach 1971, J. South Afr. Bot. 37(1): 46-51, ills..
Distribution: Ethiopia.

A. tugenensis L.E.Newton & Lavranos 1990, Cact. Succ. J. (US) 62(5): 215-217, ills..
Distribution: Kenya.

A. turkanensis Christian 1942, J. South Afr. Bot. 8(2): 173-174, t. 6.
Distribution: Kenya, Uganda.

A. tweediae Christian 1942, J. South Afr. Bot. 8(2): 175-176, t. 7.
Distribution: Kenya, Sudan (the), Uganda.

A. ukambensis Reynolds 1956, J. South Afr. Bot. 22(1): 33-35.
Distribution: Kenya.

A. umfoloziensis Reynolds 1937, J. South Afr. Bot. 3(1): 42-45, t. 2.
Distribution: South Africa (KwaZulu-Natal).

A. vacillans Forssk. 1775, Fl. Aegypt.-Arab., 74.
Incl. *Aloe audhalica* Lavranos & D.S.Hardy 1965; **incl.** *Aloe dhalensis* Lavranos 1965.
Distribution: Yemen.

A. vallaris L.C.Leach 1974, J. South Afr. Bot. 40(2): 111-115, ills..
Distribution: Angola.

A. vanbalenii Pillans 1934, South Afr. Gard. 24: 25.
Distribution: South Africa (KwaZulu-Natal).

A. vandermerwei Reynolds 1950, Aloes South Afr. [ed. 1], 268-270, ills..
Distribution: South Africa (Northern Prov.).

A. vaombe Decorse & Poiss. 1912, Recherch. Fl. Merid. Madag., 96.
Distribution: Madagascar.

A. vaombe var. **poissonii** Decary 1921, Bull. Econ. Madag. 18: 23.
Distribution: Madagascar.

A. vaombe var. **vaombe**
Distribution: Madagascar.

A. vaotsanda Decary 1921, Bull. Econ. Madag. 18: 23.
Distribution: Madagascar.

A. variegata L. 1753, Spec. Pl. [ed. 1], 1: 321.
Incl. *Aloe punctata* Haw. 1804; **incl.** *Aloe variegata* var. *haworthii* A.Berger 1908; **incl.** *Aloe ausana* Dinter 1931.
Distribution: Namibia, South Africa (Northern Cape, Western Cape, Eastern Cape, Free State).

A. vera (L.) Burm.f. 1768, Fl. Indica, 83.
≡ *Aloe perfoliata* var. *vera* L. 1753; **incl.** *Aloe barbadensis* Mill. 1768 ≡ *Aloe perfoliata* var. *barbadensis* (Mill.) Aiton 1789; **incl.** *Aloe vulgaris* Lam. 1783; **incl.** *Aloe elongata* Murray 1789; **incl.** *Aloe flava* Pers. 1805; **incl.** *Aloe barbadensis* var. *chinensis* Haw. 1819 ≡ *Aloe chinensis* (Haw.) Baker 1877 ≡ *Aloe vera* var. *chinensis* (Haw.) A.Berger 1908; **incl.** *Aloe indica* Royle 1839; **incl.** *Aloe vera* var. *littoralis* K.D.Koenig *ex* Baker 1880; **incl.** *Aloe lanzae* Tod. 1891 ≡ *Aloe vera* var. *lanzae* (Tod.) A.Berger 1908; **incl.** *Aloe vera* var. *wratislaviensis* Kostecka-Madalska 1953.
Distribution: Origin uncertain, probably Arabia.

A. verecunda Pole-Evans 1917, Trans. Roy. Soc. South Afr. 5: 703.
Distribution: South Africa (North-West Prov., Northern Prov., Gauteng, Mpumalanga).

A. versicolor Guillaumin 1950, Bull. Mus. Nation. Hist. Nat., Sér. 2, 21: 723.
Distribution: Madagascar.

A. veseyi Reynolds 1959, J. South Afr. Bot. 25(4): 315-317, pl. 32.
Distribution: Tanzania (United Republic of), Zambia.

A. viguieri H.Perrier 1927, Bull. Acad. Malgache, n.s. 10: 20.
Distribution: Madagascar.

A. viridiflora Reynolds 1937, J. South Afr. Bot. 3(4): 143-145, t. 23.
Distribution: Namibia.

A. vituensis Baker 1898, in Thiselton-Dyer & al., Fl. Trop. Afr. 7: 458.
Distribution: Kenya, Sudan (the).

A. vogtsii Reynolds 1936, J. South Afr. Bot. 2(3): 118-120, t. 15.
Distribution: South Africa (Northern Prov.).

A. volkensii Engl. 1895, Pfl.-welt Ost-Afr., Teil C, 141.
Distribution: Kenya, Tanzania (United Republic of), Uganda.

A. volkensii ssp. **multicaulis** S.Carter & L.E.Newton 1994, Fl. Trop. East. Afr., Aloaceae, 56.
Distribution: Kenya, Tanzania (United Republic of), Uganda.

A. volkensii ssp. **volkensii**
Incl. *Aloe stuhlmannii* Baker 1898.
Distribution: Kenya, Tanzania (United Republic of).

A. vossii Reynolds 1936, J. South Afr. Bot. 2(2): 65-68, t. 4.
Distribution: South Africa (Northern Prov.).

A. vryheidensis Groenew. 1937, Tydskr. Wetensk. Kuns 15: 129-131.
Incl. *Aloe dolomitica* Groenew. 1938.
Distribution: South Africa (KwaZulu-Natal).

A. whitcombei Lavranos 1995, Cact. Succ. J. (US) 67(1): 30-33, ills..
Distribution: Oman.

A. wildii (Reynolds) Reynolds 1964, Kirkia 4: 13
≡ *Aloe torrei* var. *wildii* Reynolds 1961.
Distribution: Zimbabwe.

A. wilsonii Reynolds 1956, J. South Afr. Bot. 22(3): 137-140.
Distribution: Kenya, Uganda.

A. wollastonii Rendle 1908, J. Linn. Soc. Bot. 38: 238.
Incl. *Aloe angiensis* De Wild. 1921; **incl.** *Aloe bequaertii* De Wild. 1921; **incl.** *Aloe lanuriensis* De Wild. 1921; **incl.***Aloe angiensis* var. *kitaliensis* Reynolds 1955 ≡ *Aloe lateritia* var. *kitaliensis* (Reynolds) Reynolds 1966.
Distribution: Democratic Republic of the Congo Kenya, Tanzania (United Republic of), Uganda.

A. wrefordii Reynolds 1956, J. South Afr. Bot. 22(3): 141-143.
Distribution: Kenya, Sudan (the), Uganda.

A. yavellana Reynolds 1954, J. South Afr. Bot. 20(1): 28-30, t. 4.
Distribution: Ethiopia.

A. yemenica J.R.I.Wood 1983, Kew Bull. 38(1): 20-21, ills..
Distribution: Yemen.

A. zebrina Baker 1878, Trans. Linn. Soc. London, Bot. 1: 264.
  Incl. *Aloe platyphylla* Baker 1878; incl. *Aloe transvaalensis* Kuntze 1898; incl. *Aloe lugardiana* Baker 1901; incl.*Aloe bamangwatensis* Schönland 1904; incl. *Aloe ammophila* Reynolds 1936; incl. *Aloe laxissima* Reynolds 1936; incl. *Aloe angustifolia* Groenew. 1938 (*nom. illeg.*, Art. 53.1); incl. *Aloe transvaalensis* var. *stenacantha* F.S.Mull. 1940.
  Distribution: Angola, Botswana, Mozambique, Malawi, Namibia, South Africa (Gauteng, Mpumalanga, Northern Prov.), Zambia, Zimbabwe.

A. zombitsiensis Rauh & M.Teissier 2000, in Kakt. And. Sukk., 51(8): 201-203.
  Distribution: Madagascar (SW)

PART IV: ACCEPTED TAXA - PACHYPODIUM
Further details on homotypic names, basionyms and heterotypic synonyms

QUATRIEME PARTIE: TAXONS ACCEPTES - PACHYPODIUM
Autres indications sur les noms homotypiques, les basionymes et les synonymes
hétérotypiques

PARTE IV: TAXA ACEPTADOS - PACHYPODIUM
Mayor información sobre nombres homotípicos, basionímicos y sinónimos heterotípicos

P. ambongense Poiss. 1924, Bull. Acad. Malgache, n.s. 6: 162, t. 5.
Distribution: Madagascar (Ambongo).

P. baronii Costantin & Bois 1907, Ann. Sci. Nat. Bot., Sér. 9, 6: 317-318.
Distribution: Madagascar (N).

P. baronii var. **baronii**
Incl. *Pachypodium baronii* var. *erythreum* Poiss. 1924; **incl.** *Pachypodium baronii* var.
typicum Pichon 1949 (*nom. nud.*, Art. 24.3).
Distribution: Madagascar (N).

P. baronii var. **windsorii** (Poiss.) Pichon 1949, Mém. Inst. Sci. Madagascar, Sér. B, Biol. Vég.
2(1): 123.
≡ *Pachypodium windsorii* Poiss. 1917.
Distribution: Madagascar (N) (Windsor Castle).

P. bispinosum (L.) A.DC. 1844, Prodr. Syst. Regni Veg. 8: 424.
≡ *Echites bispinosa* L.f. 1781 ≡ *Belonites bispinosa* (L.f.) E.Mey. 1837; **incl.** *Pachypodium*
glabrum G.Don 1838; **incl.***Pachypodium tuberosum* var. *loddigesii* A.DC. 1844.
Distribution: South Africa (Eastern Cape).

P. brevicaule Baker 1887, J. Linn. Soc., Bot. 22: 503.
Distribution: Madagascar (C).

P. decaryi Poiss. 1916, Bull. Acad. Malgache, n.s. 3: 235-236, pl. 10.
Distribution: Madagascar (N).

P. densiflorum Baker 1886, J. Linn. Soc., Bot. 22: 503.
Distribution: Madagascar (C).

P. densiflorum var. **brevicalyx** H.Perrier 1934, Bull. Soc. Bot. France 81: 303.
≡ *Pachypodium brevicalyx* (H.Perrier) Pichon 1949; **incl.** *P. densiflorum* var. *densiflorum*
Distribution: Madagascar (C)

P. geayi Costantin & Bois 1907, Compt. Rend. Hebd. Séances Acad. Sci. 145: 269.
Distribution: Madagascar (S).

P. ×hojnyi Halda 1998, Cactaceae etc. 8(3): 83-85, ills. (**incl.** p. 81).

P. horombense Poiss. 1924, Bull. Acad. Malgache, n.s. 6: 165, t. 9.
≡ *Pachypodium rosulatum* var. *horombense* (Poiss.) G.D.Rowley 1973.
Distribution: Madagascar (C and S).

P. lamerei Drake 1899, Bull. Mus. Hist. Nat. (Paris) 5(6): 308.
≡ *Pachypodium rutenbergianum* forma *lamerei* (Drake) Poiss. 1924; **incl.** *Pachypodium*
champenoisianum Boiteau 1941; **incl.** *Pachypodium lamerei* var. *typicum* Pichon 1949 (*nom.*
nud., Art. 23.4); **incl.** *P. lamerei* var. *ramosum* (Costantin & Bois) Pichon 1949, Mém. Inst.

Sci. Madagascar, Sér. B, Biol. Vég. 2(1): 113; ≡ *Pachypodium ramosum* Costantin & Bois 1907; **incl.** *Pachypodium menabeum* Léandri 1934.
Distribution: Madagascar (S).

P. lealii Welw. 1869, Trans. Linn. Soc. London 27: 45.
Distribution: Angola, Botswana, Namibia, South Africa, Zimbabwe.

P. lealii ssp. **lealii**
Incl. *Pachypodium giganteum* Engl. 1894.
Distribution: Angola, Botswana, Namibia.

P. lealii ssp. **saundersii** (N.E.Br.) G.D.Rowley 1973, Nation. Cact. Succ. J. 28(1): 4.
≡ *Pachypodium saundersii* N.E.Br. 1892.
Distribution: South Africa (to N KwaZulu-Natal), Zimbabwe (S).

P. namaquanum (Wiley *ex* Harv.) Welw. 1869, Trans. Linn. Soc. London 27: 45.
≡ *Adenium namaquanum* Wiley *ex* Harv. 1863.
Distribution: Namibia (S), South Africa (Northern Cape: Namaqualand).

P. ×rauhii Halda 1997, Cactaceae etc. 7(3): 108-111, ills..
Distribution: Madagascar (SW) (Itremo).

P. rosulatum Baker 1882, J. Linn. Soc., Bot. 20: 219.
Distribution: Madagascar.

P. rosulatum forma **bicolor** (Lavranos & Rapan.) G.D.Rowley 1998, Bradleya 16: 107, in clavi.
≡ *Pachypodium bicolor* Lavranos & Rapan. 1997.
Distribution: Madagascar (W) (along Tsiribihina River).

P. rosulatum var. **eburneum** (Lavranos & Rapan.) G.D.Rowley 1998, Bradleya 16: 107, in clavi.
≡ *Pachypodium eburneum* Lavranos & Rapan. 1997.
Distribution: Madagascar (C plateau, Mt. Ibity).

P. rosulatum var. **gracilius** H.Perrier 1934, Bull. Soc. Bot. France 81: 306.
Distribution: Madagascar (Isalo Mts.).

P. rosulatum var. **inopinatum** (Lavranos) G.D.Rowley 1998, Bradleya 16: 107, in clavi.
≡ *Pachypodium inopinatum* Lavranos 1996.
Distribution: Madagascar (C).

P. rosulatum var. **rosulatum**
Incl. *Pachypodium cactipes* K.Schum. 1895; **incl.** *Pachypodium drakei* Costantin & Bois 1907 ≡ *Pachypodium rosulatum* var. *drakei* (Costantin & Bois) Markgr. 1976; **Incl.** *Pachypodium rosulatum* var. *stenanthum* Costantin & Bois; **incl.** *Pachypodium rosulatum* var. *typicum* Costantin & Bois 1907 (*nom. nud.*, Art. 24.3); **incl.** *Pachypodium rosulatum* var. *delphinense* H.Perrier 1934.
Distribution: Madagascar.

P. rutenbergianum Vatke 1885, Abh. Naturwiss. Vereine Bremen 9: 125.
Distribution: Madagascar.

P. rutenbergianum var. **meridionale** H.Perrier 1934, Bull. Soc. Bot. France 81: 311.
≡ *Pachypodium meridionale* (H.Perrier) Pichon 1949.
Distribution: Madagascar (SW).

P. rutenbergianum var. **rutenbergianum**
Incl. *Pachypodium rutenbergianum* var. *typicum* H.Perrier 1934 (*nom. nud.*, Art. 24.3).
Distribution: Madagascar (NW).

P. rutenbergianum var. **sofiense** Poiss. 1924, Bull. Acad. Malgache, n.s. 6: 3.
  Incl. *Pachypodium rutenbergianum* var. *perrieri* Poiss. 1924 ≡ *Pachypodium sofiense* (Poiss.)
  H.Perrier 1934.
  Distribution: Madagascar (NW).

P. succulentum (L.f.) Sweet 1830, Hort. Brit., ed. 2, 594.
  ≡ *Echites succulenta* L.f. 1781 ≡ *Belonites succulenta* (L.f.) E.Mey. 1837; **incl.** *Pachypodium*
  tubersosum Lindley 1830; **incl.** *Pachypodium tomentosum* G.Don 1838; **incl.** *Pachypodium*
  griquense L.Bolus 1932; **incl.** *Pachypodium jasminiflorum* L. Bolus 1932.
  Distribution: South Africa (Northern Cape, Western Cape, N-wards to Free State).